高等院校土建学科双语教材（中英文对照）

◆ 工程管理专业 ◆

BASICS

施工进度计划

CONSTRUCTION SCHEDULING

［德］伯特·比勒费尔德 编著

杨 璐 柳美玉 译

中国建筑工业出版社

著作权合同登记图字：01－2011－0782 号

图书在版编目（CIP）数据

施工进度计划／（德）比勒费尔德编著；杨璐，柳美玉译．—北京：中国建筑工业出版社，2011.6
高等院校土建学科双语教材（中英文对照）◆工程管理专业◆
ISBN 978－7－112－12957－7

Ⅰ.①施… Ⅱ.①比…②杨…③柳… Ⅲ.①施工进度计划－高等学校－教材－汉、英 Ⅳ.①TU722

中国版本图书馆 CIP 数据核字（2011）第 026940 号

责任编辑：孙　炼
责任设计：陈　旭
责任校对：陈晶晶　姜小莲

高等院校土建学科双语教材（中英文对照）
◆工程管理专业◆
施工进度计划
［德］伯特·比勒费尔德　编著
杨　璐　柳美玉　译
*
中国建筑工业出版社出版、发行（北京西郊百万庄）
各地新华书店、建筑书店经销
北京嘉泰利德公司制版
北京云浩印刷有限责任公司印刷
*
开本：880×1230 毫米　1/32　印张：5　字数：140 千字
2011 年 7 月第一版　　2011 年 7 月第一次印刷
定价：**19.00** 元
ISBN 978－7－112－12957－7
（20196）

中文部分目录

CONTENTS

序

将进度计划和施工过程协调统一是一项非常复杂的任务，特别对于规模较大的工程，其中涉及大量不同的工作。由于不同工作部门之间、不同工种承包商之间以及二者相互之间的联系和影响日益增多，需要协调众多的工作者以及工作内容之间的相互关系。对于工程设计人员，制定施工进度计划是把握控制整个工程最重要的方法。施工进度计划向各个参与施工的公司明确了合同中各工种的时间期限，同时也是用来主动应对施工过程中可能遇到的偶然事件和各种干扰的重要工具。

由于大学毕业生和初级专业人员在工程经验方面比较缺乏，在着手进行第一个工程的时候经常不知道该如何管理整个设计工程和施工过程中的参与者。经常会面临以下一些问题：哪些工作是必须进行协调安排的？不同施工过程的施工顺序如何确定？这些施工过程将持续多长时间？本书面向的读者正是相关专业的学生和初级专业人员。该书通过循序渐进的方法并结合实际工程，对施工进度计划的制订、施工设计和施工过程在其中的表达方法以及如何在实际工程中使用施工进度计划进行了讲解。

编辑　伯特·比勒费尔德（Bert Bielefeld）

FOREWORD

Coordinating the planning and construction process is a complex task, which involves a great deal of responsibility, especially in larger building projects. Due both to the increasing interconnection between components and to the specialization of contractors, a large number of participants and their work must be organized. Scheduling is the most important means by which project planners control the entire process. It forms the basis of the contractual deadlines given to the participating construction companies, and is also a tool that is used actively to respond to unforeseen happenings and disruptions during the planning and construction process.

Due to their lack of experience, university graduates and entry-level professionals embarking on their initial projects are often unsure of how to manage the participants in the planning and construction process. Typical questions are: What work must be coordinated? What is the sequence of work steps? And how long do these steps last? *Basics Construction Scheduling* is directed at students and entry-level professionals at this early stage of the game. In a step-by-step, practical way, it shows how a schedule is created, how it represents the planning and construction process, and how it can be used as a tool in the real world.

Bert Bielefeld, Editor

INTRODUCTION

The translation of an initial idea into a completed building is a lengthy and extremely complex undertaking. The large number of people involved—construction contractors, planners and owner-builders—make it necessary to coordinate all the different contributions to the process closely.

Architects and project planners represent the owner in technical matters and must work to ensure that the entire process runs as smoothly as possible. Looking after the owner's interests, they coordinate all the participants in the planning process and monitor the contractors on the construction site. Larger projects often entail twenty to thirty participants or more in the planning and construction processes, which results in complex links and interdependencies. The various participants are often unable to understand or judge how their specific work is linked to the project workflows as a whole. As a result, architects have a special coordinating responsibility since their planning encompasses the entire range of specialized tasks involved in a project, and they are therefore the only participants in the process who have the "big picture."

Scheduling is a tool that is used in all stages of this process. The present book explains its foundations and applications, addressing all forms and depths of representation and providing practical information on typical processes. Its goal is to give students a quick, real-world introduction to the material. Yet coordination work is not over once a schedule has been created. It is a work process that must be constantly updated and made more precise. A good deal of preliminary consideration and refinement of detail is required in order to specify the phases into which work on a site is ultimately organized. The following chapters describe which participants and steps need to be taken into account when creating a schedule.

CREATING A SCHEDULE

SCHEDULE ELEMENTS

To begin with, a description of a few key terms and the various elements of a schedule is in order.

Period and deadline

Planners distinguish between a deadline and a time period. The word deadline describes a specific point in time, such as the day on which part of a project must be completed, while a period is a span of time (e.g. completion of a job within fourteen days).

Tasks

Tasks are the very foundation of the schedule and refer to self-contained work units (e.g. tiling the ground floor). If several tasks are combined (e.g. tiling and plastering), the result is a summary task › **Chapter Creating a schedule, The structure of the project schedule**

Planning the duration and sequence of tasks

Task duration is the time needed to complete a task. It is a factor of production quantity and productivity. › **Chapter Creating a schedule, Planning task duration**

The calculation of duration is referred to as duration planning. Establishing the dependencies between activities is referred to as sequence planning. Taken together, duration and sequence planning form the basis of construction scheduling. › **Fig. 1**

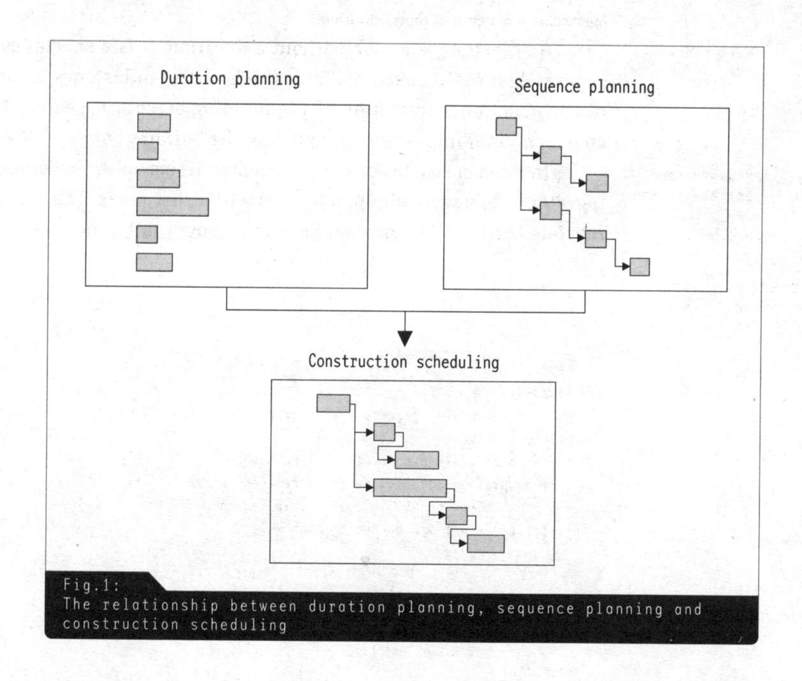

Fig.1:
The relationship between duration planning, sequence planning and construction scheduling

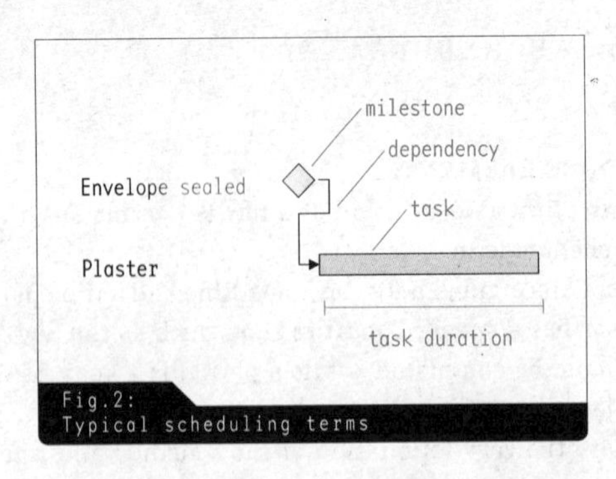

Fig.2:
Typical scheduling terms

Construction methods and resources

Construction method refers to the technical procedure for carrying out a task.

The equipment and labor necessary to perform a task are called resources. While preparing for a building project, construction companies plan their resources in order to calculate costs precisely and define construction methods. The result serves as the foundation for their bid. While calculating resources has only limited significance for the architect doing the scheduling, smooth workflow requires a realistic assessment of task durations, for which resource planning provides a foundation. › Chapter Creating a schedule, Depth of representation

Milestones

A milestone is a task without a duration. It is a special event entered separately into the schedule. Typical scheduling milestones include the start of construction, completion of the building structure, sealing the building envelope, final inspection and putting the building into operation. › Fig. 2

Dependencies between tasks

In most cases, tasks are not isolated items on the schedule but are integrated into a web of dependencies with other tasks. There can be several reasons for this. The normal case is a sequential dependency: task B can

\\Example:
In any construction process there may be many ways to achieve the desired results. For instance, a reinforced concrete ceiling may be built of prefabricated elements or cast on site. Wall tiles can be laid in a thin bed on plaster or in a thick bed on a rough wall.

only begin once task A is finished (e.g. ground-floor walls → ground-floor ceiling → upper-floor walls). ›

That said, some tasks can only be performed jointly in a parallel process (e.g. setting up scaffolding floor by floor as the structure of a multistory building goes up). Often these process dependencies can be broken down into sequential dependencies by using a higher level of detail.

By contrast, it is often impossible for finishing contractors to work in parallel during a number of construction phases (e.g. screed and plastering work). In this case we speak of one task <u>interfering</u> with another. This is why it is essential for planners to examine mutual dependencies between specialized tasks and, if necessary, to divide the project into optimal construction phases. › **Chapter Creating a schedule, Planning task sequence and Chapter Workflows in the planning and construction process**

Types of relationships

› 

Various types of relationships play an important role in the graphic representation of dependencies between two tasks. Construction scheduling distinguishes between four types: › **Fig. 3**

- <u>Finish-to-start</u>: Task B can only begin after task A is finished. This is the most common type of relationship and may apply to activities such as the construction of interior walls (A) and interior plastering (B).
- <u>Finish-to-finish</u>: Task A and task B must be completed by the same time. This type of relationship exists when tasks A and B provide the foundation for an additional task. Examples are installing windows (A) and sealing the roof (B), which create the airtight building envelope necessary for interior work.
- <u>Start-to-finish</u>: Task B must end when task A begins. In this type of relationship, one task can be scheduled at the latest possible point in time before it interferes with another task.

\\Note:
If sequential dependencies are not examined in detail during scheduling work, the result can often be disruptions and delays in the construction process. If, for instance, a disabled-accessible, steel-framed door needs to be installed, the required electrical outlets must be installed before plastering is done. Overlooking such dependencies may result in further work being required on finished surfaces.

\\Tip:
The most popular construction scheduling programs support the types of relationships described above. As a rule, they assign each task its own number, which can be used to denote dependencies. For example, if a task needs to begin after task no. 5, the previous task is marked 5FTS, where FTS stands for a finish-to-start relationship.

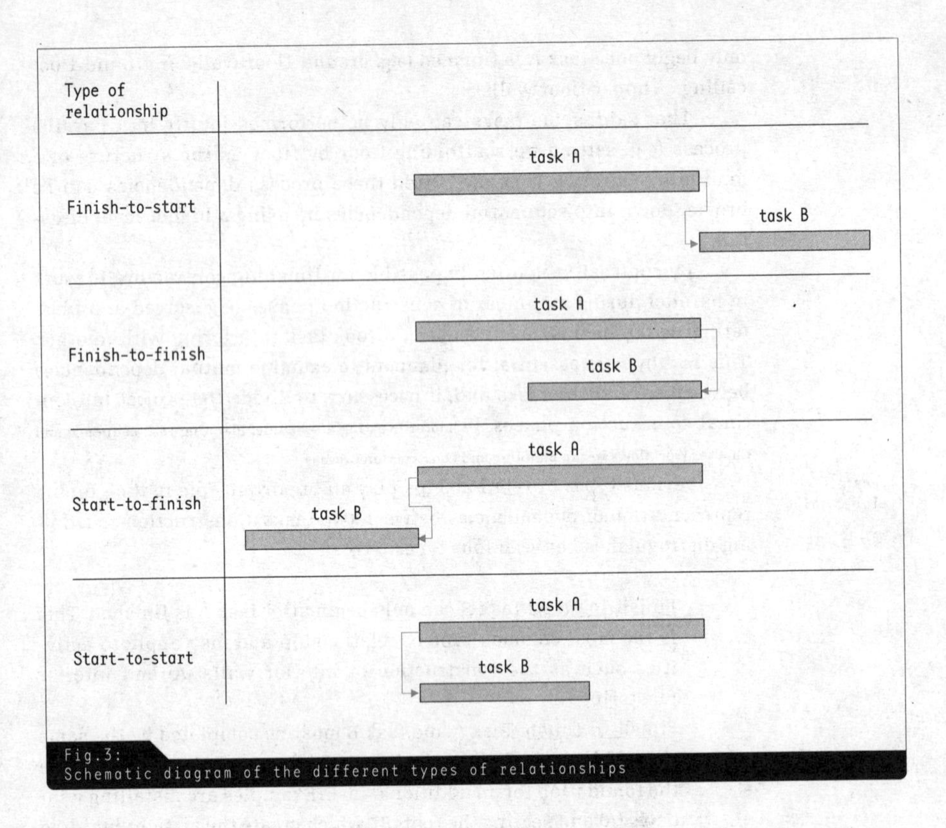

Fig. 3:
Schematic diagram of the different types of relationships

- Start-to-start: Task A and task B must start at the same time. This makes sense if the work can be performed in parallel—if, for instance, workers from one trade can use a crane that is operated by another contractor to deliver large building elements.

FORMS OF REPRESENTATION

There are different ways to represent a schedule graphically. The following forms of representation are used to communicate schedule contents in a clear and useful manner, depending on the goal and purpose of the schedule: › Fig. 4

_ Bar chart
_ Line diagram
_ Network diagram
_ Deadline list

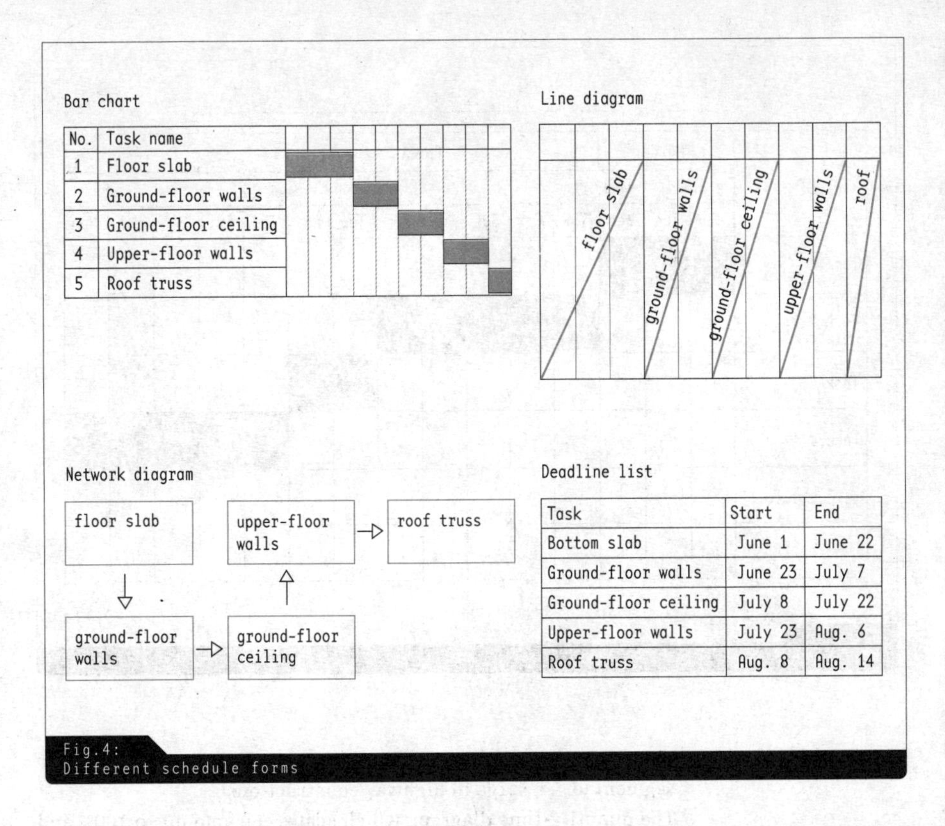

Task	Start	End
Bottom slab	June 1	June 22
Ground-floor walls	June 23	July 7
Ground-floor ceiling	July 8	July 22
Upper-floor walls	July 23	Aug. 6
Roof truss	Aug. 8	Aug. 14

Fig.4:
Different schedule forms

Bar chart

In building construction, schedules are normally shown as bar charts, also called Gantt diagrams. Time is charted along the horizontal axis, and the various tasks are listed along the vertical axis. The duration of each task is recorded as a horizontal bar along the corresponding time axis. › Fig. 5

The time axis can be divided into months, weeks and days, depending on project scope and the required degree of detail. In addition to graphically representing activities, this form of schedule commonly incorporates related information in the left-hand column to facilitate easy reading (e.g. task description, starting date, completion date, duration and, if necessary, dependencies with other tasks). Dependencies are often shown graphically as arrows between different bars.

Line diagram

The basic structure of a line diagram differs from that of a bar chart in that the units executed are shown on the second axis next to the timeline. The tasks are depicted by lines in the coordinate system. The following types of line diagrams are commonly used in the construction industry:

time axis →

	March			April										
	week 11							week 12						
	Mon	Tue	Wed	Thu	Fri	Sat	Sun	Mon	Tue	Wed	Thu	Fri	Sat	Sun	Mon
Task A															
Task B															
Task C															
Task D															
....															

Fig.5:
Bar chart principle

- The space-time diagram, which shows the quantity as a geometric segment (e.g. a stage in highway construction);
- The quantity-time diagram, which scales the amount to 100% and shows what percentage of the task has been completed, independent of the actual quantity.

The line diagram is generally used less frequently in building construction than in areas involving linear construction sites such as streets, tunnels and sewage systems, in which the construction steps follow each other in regular cycles. In building construction, many tasks, particularly those done by the finishing trades, must be performed in parallel and cannot be clearly illustrated on a line diagram. Nevertheless, the line diagram does have the advantage of being able to represent target-performance comparisons more clearly. › Fig. 6

Network diagram

A network diagram represents scheduled tasks as networks rather than as items along a time axis. Using this form of schedule, planners can effectively map out the reciprocal links between tasks, but it provides only a limited sequential overview of the entire process.

There are three types of network diagrams; the activity-on-node network is the most popular:

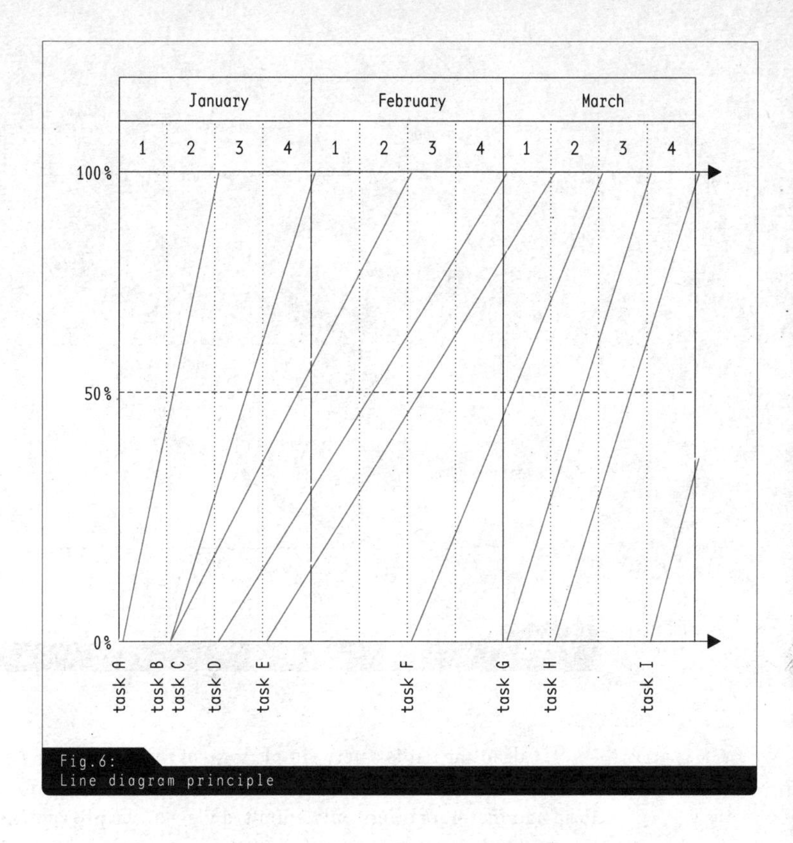

Fig.6:
Line diagram principle

_ Activity-on-node: Activities are represented by nodes and dependencies by arrows.
_ Activity-on-arrow: Activities are represented by arrows and dependencies by the links between the nodes.
_ Event-on-node: Arrows symbolize the dependencies and nodes represent the results (without durations). › Fig. 7

Network diagrams are rarely used to represent construction schedules, but they are often featured in premium construction scheduling software as an additional means of displaying bar charts. As such, they fulfill an important function in schedule creation. Since they allow for more enhanced graphic representation of dependencies between tasks than a bar chart, planning specialists can switch between these two display modes to better orient themselves within the schedule. Tasks can be recorded and assigned durations using bar charts, while mutual dependencies can be checked and represented using network diagrams.

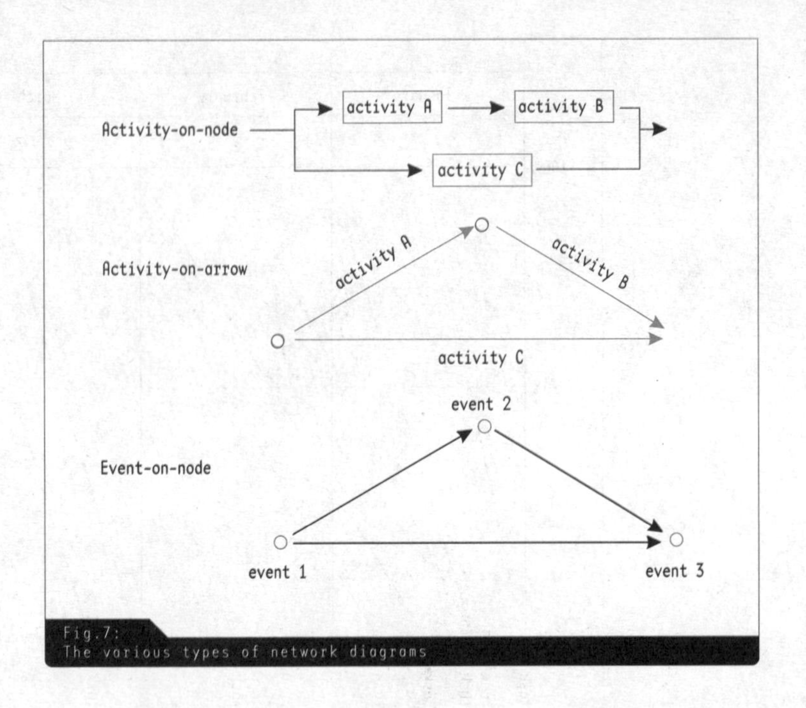

Fig.7:
The various types of network diagrams

Deadline list

A deadline list is a very simple form of representation. As used in construction scheduling, it presents important deadlines and periods in table form and therefore offers only a limited view of the project. The more deadlines the list contains, the harder it is to read and the more difficult it is for project participants to grasp the interconnections between the individual deadlines.

Deadline lists are often excerpted from the schedule for the different participants in the planning and construction process in order to inform these parties of important deadlines and time periods. Such information may relate, for instance, to the time needed by planners or experts to do their work or to the contractual specifications of the individual contractors. Lists including construction deadlines are often submitted along with tenders, and they may afterwards be incorporated into the construction contract as contractual deadlines. Many scheduling programs can output deadline lists in separate files based on construction schedules.

DEPTH OF REPRESENTATION

A schedule should always meet the specific requirements of the project concerning clarity, functionality and degree of detail. These requirements may vary widely depending on the perspective taken on the

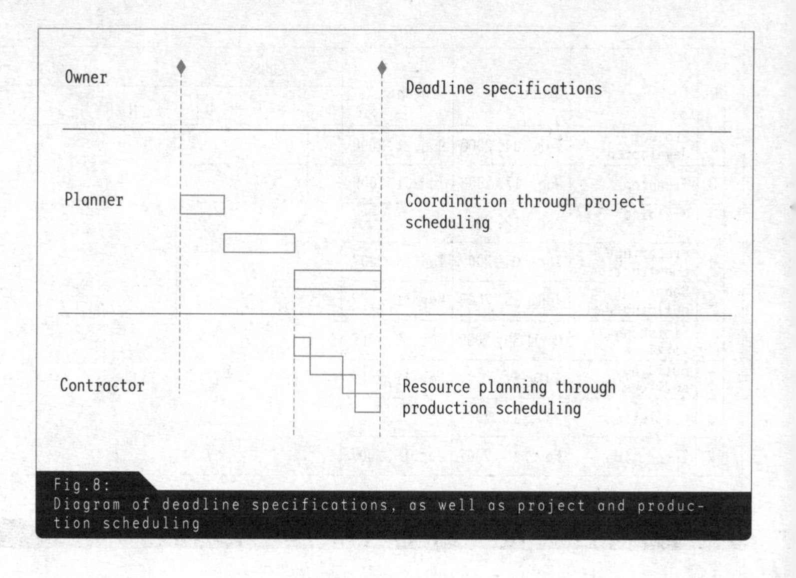

Fig. 8:
Diagram of deadline specifications, as well as project and production scheduling

construction project. There are generally three levels of detail in construction processes, all of which reflect this perspective:

_ The owner's perspective: establishing deadlines with the help of framework scheduling
_ The planner's perspective: coordinating participants with the help of project scheduling
_ The construction contractor's perspective: preparing work and planning resources with the help of production scheduling › Fig. 8

Furthermore, schedules are also categorized in terms of their timeframe (short-term, medium-term, long-term), the person creating them (owner or contractor), and their level of detail (rough, detailed, highly detailed).

Framework schedule

Owners usually have a clear idea of when they want or must begin using a building. A department store may need to be completed by the next Christmas season, or the owner may already have given notice on his or her lease and must be out by a certain date. Planners must take into account the deadlines specified by the owner. Often the financial backers of a project (banks) are also the source of deadline constraints. In order to check the owner's ideas about deadlines and roughly structure the entire process, planners create a framework schedule as an initial overview. It contains the broader sets of tasks and the intermediate deadlines of planning and construction. Typical tasks are: › Fig. 9

No.	Task	Start	Finish	2008						2009											
				J	A	S	O	N	D	J	F	M	A	M	J	J	A	S	O	N	D
1	Project development	July 04, 2008	Aug. 11, 2008																		
2	Planning	Aug. 17, 2008	Jan. 01, 2009																		
3	Building permit	Dec. 12, 2008																			
4	Preparing construction	Jan. 01, 2009	Jan. 15, 2009																		
5	Constructing building shell	Jan. 15, 2009	May 31, 2009																		
6	Finishing work	April 30, 2009	Oct. 30, 2009																		
7	Building services	July 31, 2009	Nov. 11, 2009																		
8	Completion	Nov. 11, 2009																			
9	Inspection	Nov. 15, 2009	Dec. 15, 2009																		

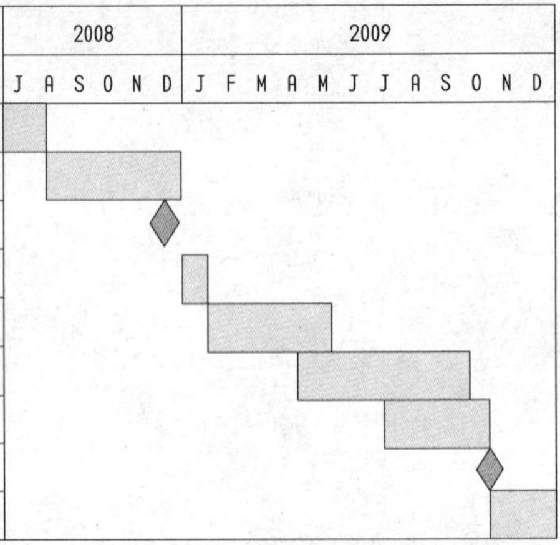

Fig.9:
Example of a framework schedule

- Project preparation
- Design
- Building permit application
- Preparing for construction work
- Start of construction work
- Building structure
- Building envelope
- Various finishing jobs
- Completion

Project schedule

The project schedule is usually created by the architect, and its objective is to coordinate all the participants in the planning and construction of a building. In order to link the different activities and effectively coordinate the work, the project schedule must have a higher level of detail than the framework schedule. › Fig. 10 The most important factor in combining or separating out tasks is their interdependence. For instance, the building shell often requires a low level of detail since the related tasks are not carried out by different trade contractors and only serve the purpose of deadline control. By contrast, the construction of a drywall may require electrical installations, sanitary installations, door assembly and painting work, which make it necessary to break down the task into several steps. › Chapter Workflows in the planning and construction process, Finishing work As mentioned above, the most effective level of detail in a project schedule generally depends on the project's complexity and timeframe.

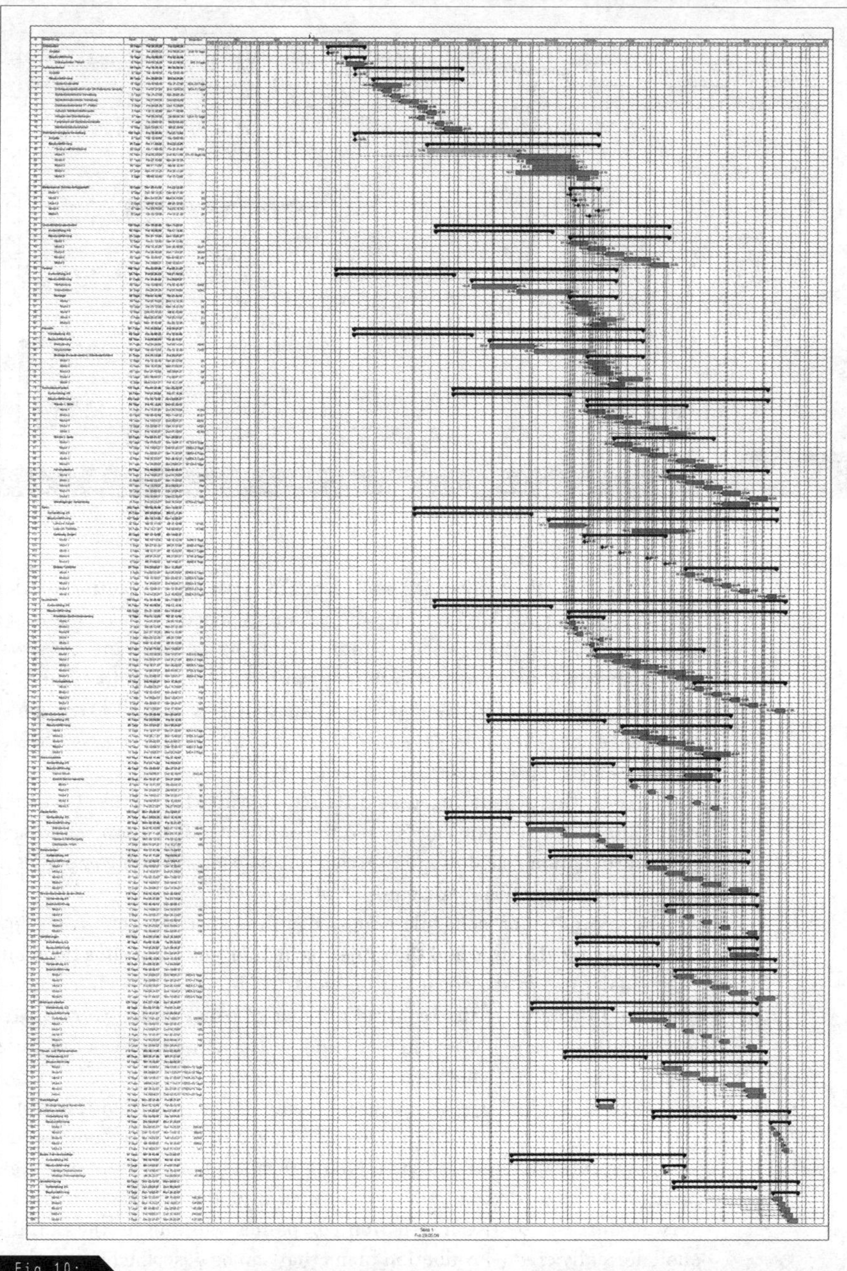

Fig.10:
Sample project schedule

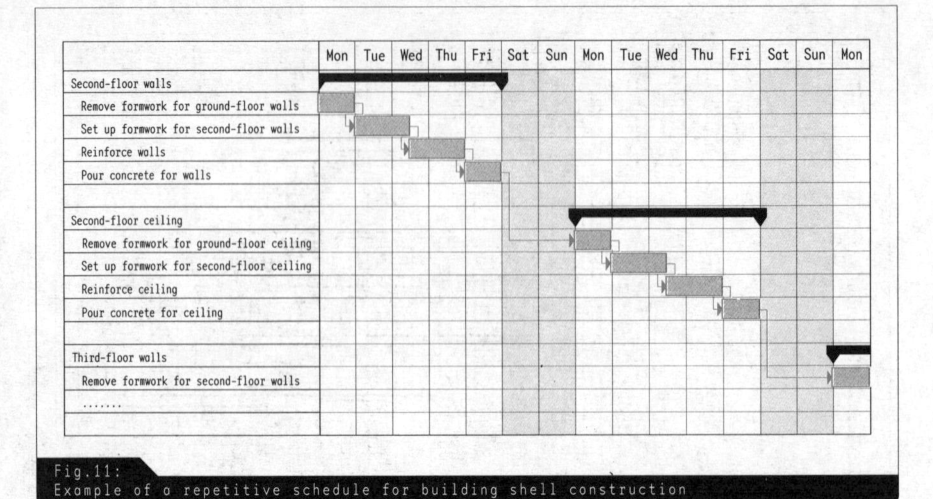

Fig.11:
Example of a repetitive schedule for building shell construction

Moreover, the project schedule should cover the execution of construction work and also integrate the planning phase so that it covers all interfaces. Ideally, the project schedule should be designed in such a way that the various users at the architectural office (construction planning, tendering, construction management) can concisely and clearly display those deadlines that are relevant to their work. › **Chapter Workflows in the planning and construction process, Coordinating planning**

Production scheduling

Production scheduling has a different objective from project scheduling. While project scheduling is used to coordinate all the project participants, production scheduling is carried out by construction companies to plan personnel, material and equipment resources.

The production schedule adopts deadline specifications from both the framework schedule and the project schedule, and translates them into individual steps in the construction work. This form of schedule enables planners to determine the required number of construction workers and to promptly allocate both equipment and sufficient quantities of material in order to avoid bottlenecks.

The task which the construction company adopts from the project schedule (e.g. a concrete ceiling installed over the ground floor) is broken down into individual steps (placing forms, adding reinforcing steel, pouring concrete, drying, removing forms). It is also assigned the necessary resources. › Fig. 11 Construction companies working on the building shell normally create production plans that can be described as repetitive schedules due to the cyclical nature of the individual work steps. In this case, the construction project is divided into several identical construction phases. Since quantities are identical, companies only need to develop a production schedule for the steps in a single phase and apply it to additional cycles.

Production schedules are used less often in the finishing trades since there are many links between the specialized tasks that place limits on the finishing contractors' ability to organize and schedule the work themselves.

CREATING A FRAMEWORK SCHEDULE

The usual objective of the framework schedule is to examine the feasibility of the owner's deadlines and integrate participants roughly into a plan. When project managers or professional owners participate in larger projects, the framework schedule is often created by the owner and given to planners as a set of parameters.

Deadline specifications from the owner

Depending on the owner's preferences, deadline specifications will take the form of either a directly specified completion deadline or indirectly defined deadlines and periods that participants must observe (e.g. the start of construction in the current fiscal year to give tax advantages). The time span from the start of planning to the completion of the building marks out the framework for the entire planning and construction period.

Dividing the project into planning and construction

Dividing the project into planning and construction periods is a crucial step that allows planners to examine whether both parts of the project can be implemented. While the construction period can usually be streamlined using a cushion and it is also possible to optimize the planning period time-wise, there are limits to such optimization efforts. The two phases must meet basic requirements, and related deadlines can be missed by only a narrow margin.

The start of construction is determined by the availability of the building permit, the related lead times for planning and approval, and the contractual agreement with the construction company. This agreement usually presupposes construction drawings, a call for bids, and the submission of bids from construction companies.

Construction work cannot bypass the typical sequences of the construction process to any significant extent, and there are also limits to the ways it can be organized in parallel sequences.

If a deadline is specified by the owner, the necessary construction period should be calculated backwards from it, or it should be estimated using comparable projects. This approach allows planners to determine when construction needs to start. They must then examine whether planning requirements can be met in the time remaining between the start of the project and the start of construction. Even when making comparisons with other projects, they should always keep the complexity of the project in mind.

If there are grounds for believing that a project is not viable, various alternatives must be considered. For instance, construction time can be shortened by using alternative construction methods (pre-manufacturing, prefabricated parts, materials with short drying times). If the completion deadline is still unrealistic, the problem should be discussed with the owner at an early stage.

Organizing tasks

In addition to dividing the project into a rough planning and construction period, the framework schedule covers a few central planning steps and work sections, presenting them as individual tasks or milestones. › Chapter Creating a schedule, Schedule elements It provides a general overview of the project and ensures that the participants are deployed punctually. Project scheduling generally adds a higher level of detail, but the boundaries are fluid and additionally influenced by the owner's interest in information.

THE STRUCTURE OF THE PROJECT SCHEDULE

Since the project schedule coordinates participants, its rough structure is tailored to their work. The schedule integrates each planner and construction company separately, together with the jobs they do.

Hierarchical structure based on contract work packages

Individual tasks are combined to form summary tasks in order to structure the schedule and create a schedule hierarchy. For example, individual tasks can be assigned to a component group or to a construction phase group. › Fig. 12

The highest hierarchical level should always be the respective contract work package. The term "work package" refers to the construction jobs that have been contracted out in an agreed planning or construction contract. If a work package involves several trade contractors, each must individually be subordinated to the work package. The advantage of this approach is that the hired contractors can be clearly monitored in terms of their individual deadlines and construction jobs. Furthermore, contract

\\Example:
It is much easier to evaluate the planning scope and the construction period of a simple hall as opposed to a laboratory building of the same size that will contain sophisticated technical equipment. An additional problem is that a larger number of planners must be coordinated for the laboratory, which increases the risk of disruptions.

\\Tip:
In most scheduling programs, planners can use summary tasks to structure the schedule by "inserting" tasks at lower levels. The respective lower-level task automatically becomes a summary task, the duration of which is determined by the sum of its subordinate tasks.

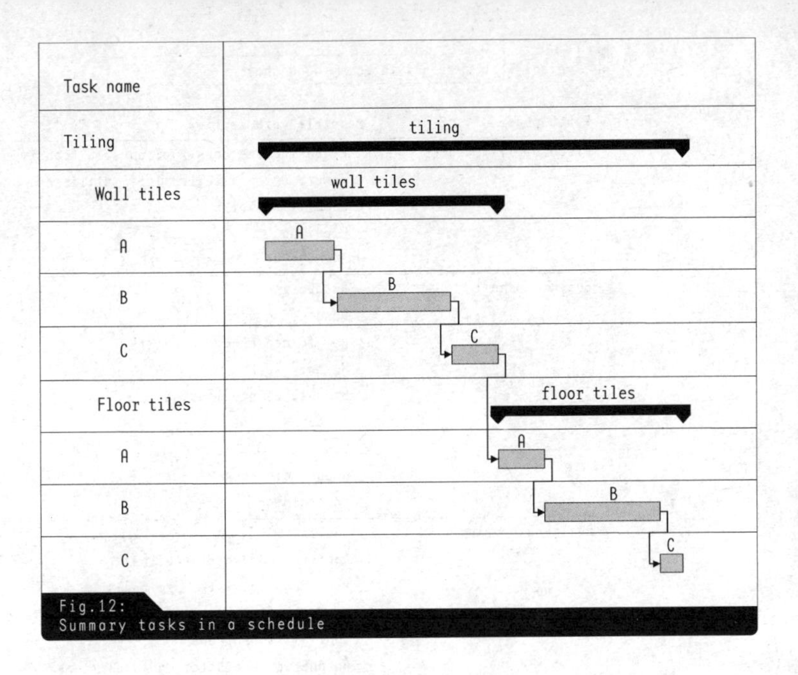

Fig.12:
Summary tasks in a schedule

award schedules can be created without additional effort and used to establish the deadlines of contract award processes. › Chapter Workflows in the planning and construction process, Coordinating construction preparation, and Chapter Working with a schedule, Updating and adjusting a schedule

Rough phases of an implementation schedule

The work packages are usually arranged in chronological order according to the progress of construction work. If the construction process is divided into segments, it may include the following broad phases:

1. Preparatory measures
2. Building shell
3. Building envelope
4. Interior finishing
5. Building services
6. Final work

Allocating tasks to the work sections

These broad chronological phases can be matched to individual work sections (sometimes with overlaps) in order to produce an initial schedule structure. Afterwards, the construction jobs are categorized under the contract work packages as sets of tasks. › Tab. 1 and Chapter Workflows in the planning and construction process

Tab.1:
Typical work sections in each rough phase

Rough phase	Possible work section
Preparatory measures	_ Preparing the construction site (construction fences, construction site trailers, utilities, etc.) _ Demolition work _ Clearing _ Excavating
Building shell	_ Excavating _ Dewatering _ Reinforced concrete work _ Masonry work _ Structural steelwork _ Timber work _ Sealing _ Ground drainage _ Scaffolding
Building envelope	_ Sealing _ Roofing/roof waterproofing _ Plumbing (rainwater drainage) _ Windows _ Shutters/sunscreens _ Facade work (plaster, natural stone, masonry, curtain wall, etc., depending on building envelope)
Finishing	_ Plastering _ Screed work _ Drywall construction _ Metalwork (e.g. railings) _ Natural/artificial stonework _ Tiling _ Parquet _ Flooring _ Painting/wallpapering
Building services	_ Ventilation systems _ Electrical work _ Sanitation/plumbing _ Heating/hot water systems _ Gas installations _ Lightning protection _ Transportation and elevator systems _ Fire protection _ Building automation _ Security technology
Final work	_ Carpentry (furniture) _ Locking systems _ Final cleaning _ Outdoor facilities

Planning task sequence and duration

The next steps involve linking the tasks with one another and calculating their duration. › Chapter Creating a schedule, Planning task sequence, and Planning task duration While dependencies in the construction process usually determine the links between tasks, planners should also take outside influences into account. For instance, they might have to observe deadlines specified by the owner or integrate events such as topping-out ceremonies that need to be scheduled before the start of summer vacations. Finally, events taking place in the area around the construction site may have an effect on the schedule (e.g. street or city festivals, utility connection dates specified by the authorities). If at all possible, planners might also find it advisable to schedule critical tasks in a period of more clement weather outside the frost season.

Dividing the project into construction phases

Dividing the work into construction phases is one of the most important steps in organizing the sequence of the different work sections and generally streamlining the schedule. Tasks such as laying floor screed are organized for the different building sections (ground-floor screed work, second-floor screed work, etc.). Smaller phases help produce overlaps between sets of tasks and are advisable since it would "straighten out" the construction process too much if, for instance, the schedule required plastering throughout the building before screed was laid. The tasks are assigned to construction phases in order to inform construction companies where they are to start and in what order the work will proceed. › Fig. 13

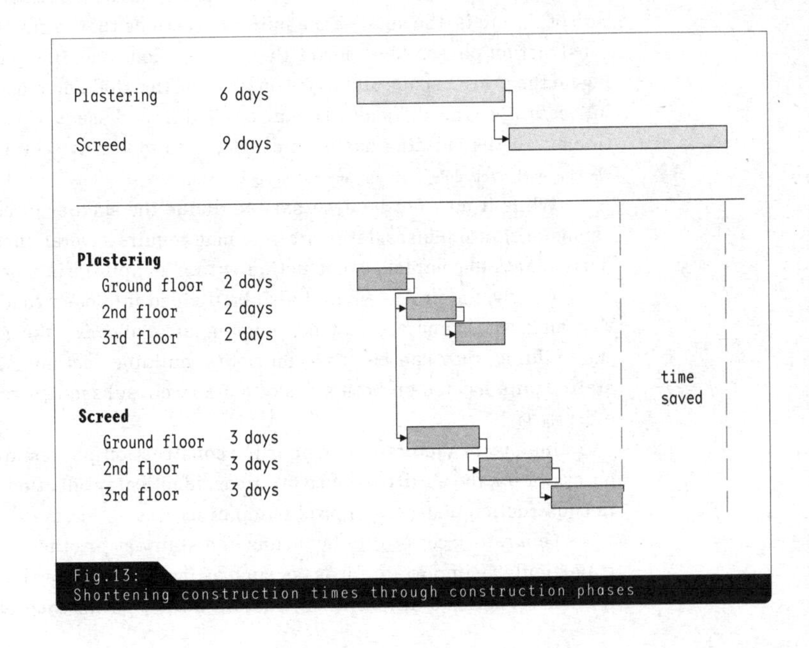

Fig. 13:
Shortening construction times through construction phases

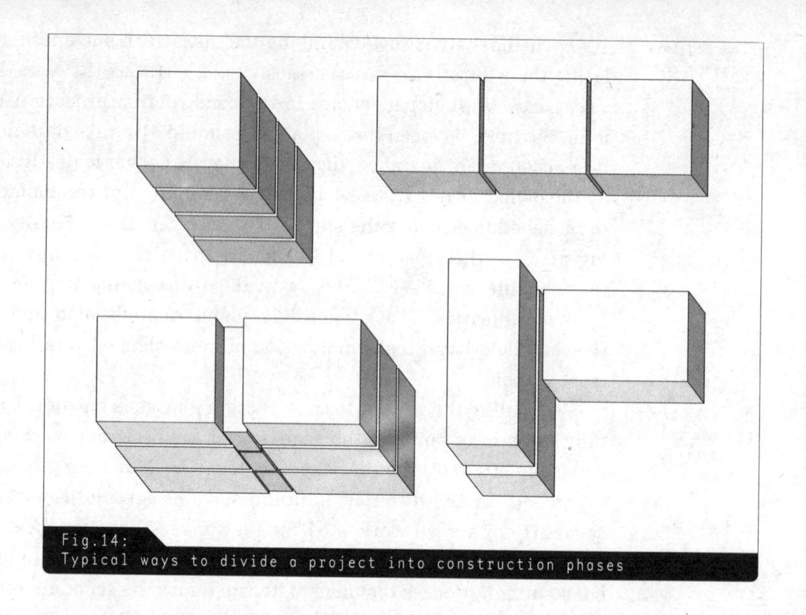

Fig.14:
Typical ways to divide a project into construction phases

Planners must consider carefully how they divide the project into construction phases when they structure the schedule since changes made to these phases at a later date can result in a great deal of extra work. The rule of thumb is, the smaller the building sections that forms the basis of construction phases, the shorter the construction time. However, depending on the project scope and time constraints, this division must not be too intricate since it is difficult to create a schedule and use it on the construction site if the building has been divided into too many sections. › **Chapter Working with a schedule, Updating and adjusting a schedule**

While it may not be necessary to divide the smaller projects (home extensions) into phases, large projects may require several phases for the participants to complete construction within an appropriate period.

Finally, the project should also be divided into construction phases that make sense and are easy to communicate. Following the geometry of the building, they can be based on floors, building sections accessed by stairs, units located on both sides of a stairwell, subsequent rental units, etc. **› Fig. 14**

Important factors to consider for construction phases are separate accessibility, the ability to close off areas, identical production quantities in construction phases, and production processes.

Separate accessibility by means of a stairway or other access route is particularly important for tasks such as laying screed or flooring since it ensures that the different contractors do not get in each other's way

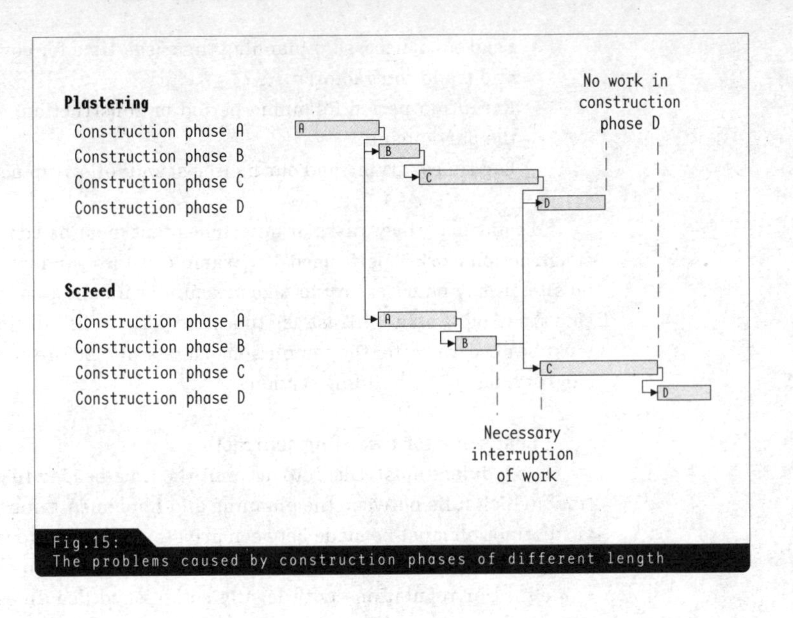

Fig.15:
The problems caused by construction phases of different length

and are not prevented from reaching their work stations by closed-off areas.

Even so, closing off areas can help to protect completed building sections and surfaces from damage. If areas are closed off or only open to the contractors currently working on them, on-site damage, dirt and theft can be limited, and the responsible party can be identified more easily.

As regards production quantities, planners should define the various construction phases so as to ensure that each phase and work section involves an identical volume of work. This creates continual cycles and avoids long waiting times for the different contractors. › Fig. 15

An additional factor to be considered when dividing a project into construction phases is the various production processes used by the individual contractors. As a rule, a building structure is built floor by floor (from bottom to top), but there are a number of building services contractors that work either along installation paths such as sewage pipes (from top to bottom) or in self-contained cycles (e.g. subdivisions in a rental unit). This is a constant source of misunderstanding and mutual disruptions.

PLANNING TASK SEQUENCE

In order to gain a better understanding of the different tasks, we will systematically follow a single project participant through the typical sequence of the work he or she performs. This sequence can usually be divided into three rough phases:

- Lead time (necessary planning time, lead time for contract awards and trade contractors)
- Execution period (planning period or construction, depending on the participant)
- Lag times (drying and curing times) and follow-up periods

Lead times cover tasks or milestones that must be scheduled before construction work is performed. For example, before windows are installed on site, it may be necessary to take measurements and plan and prefabricate the windows. In contrast, lag times are periods, like drying times, that must be observed after the completion of a task and before additional work can be done on the building section.

Lead time for awarding contracts

Schedulers must take into account the time needed to award a contract, which falls between the planning and implementation period. A basic distinction must be made between private-sector and government procedures. The government usually awards contracts on the basis of strict guidelines or regulations with legally established deadlines. In the private sector, these regulations are not binding. The contract award process can therefore be organized along less formal and more direct lines, yet it should nonetheless adhere to certain minimum periods in order to allow all participants to respond in the proper manner. The deadlines in government regulations are therefore a good foundation for the private sector.

› Schedule milestones

The contract award process involves several steps › Fig. 16 and Chapter Workflows in the planning and construction process, Coordinating construction preparation The schedule should, at the very least, include the following tasks or milestones:

- Publication, which is necessary for most government procedures
- Issuing invitations to tender: the deadline by which the planner must complete all documents

\\Note:
A lead time for awarding contracts should also be taken into account for planning services in order to find the planning specialists who are best suited and have the most experience for the job (e.g. those with expertise in fire protection for conversions of existing buildings). Public invitations to tender may also be necessary when awarding planning contracts.

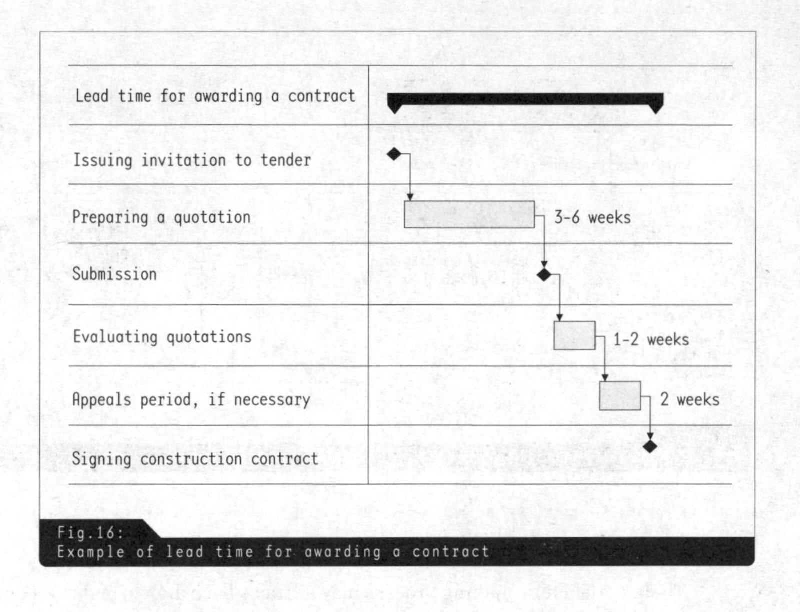

Fig.16:
Example of lead time for awarding a contract

_ Submission: the deadline by which companies must submit their tenders
_ Signing the construction contract: a deadline for the building owner
_ Start of construction

When scheduling the lead time for the award of each contract work package, planners calculate backwards since the lead time is usually scheduled to that construction work begins promptly on site.

The construction contract and the start of construction

A period of at least two weeks must be scheduled between the signing of the construction contract and the start of construction since the contracting company must first make plans before it can begin the construction process (material requirements, transportation to the site, construction site facilities, etc.).

Submission

Sufficient time must also be left between the submission of a tender and the signing of the construction contract—at least one or two weeks, depending on the complexity of the work. During this period, planners examine all the tenders and compile a price comparison list that will be used by the owner as the basis for deciding which construction company to hire. Any unclear items or deviations in the tenders must be discussed with the participants. Government procedures often stipulate an appeals period for bidders who have submitted less favorable offers. The owner's

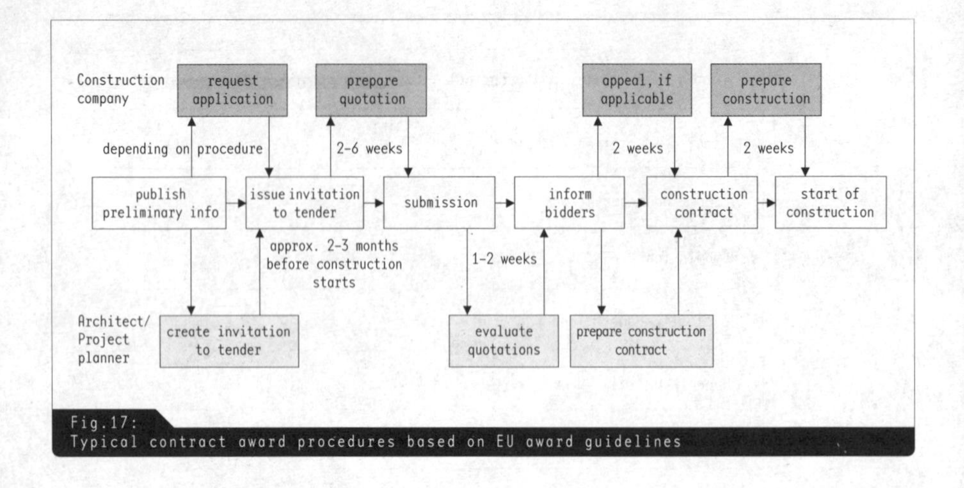

Fig.17:
Typical contract award procedures based on EU award guidelines

decision-making process may at times be arduous, making a period of more than two weeks necessary.

Sending out the call for tender

Once the planner issues the invitation to tender, the tendering construction company must put together a solid offer by the submission deadline. Depending on the complexity of the required construction work, the company may require a good deal of time and effort to calculate its offer. It must therefore be given sufficient time to do so—usually around six weeks. The reason for this is that the construction company will have to ask suppliers for their prices or create its own call for tenders in order to get information from subcontractors on sections of the construction work. Even if there is great urgency, construction companies are usually unable to put together a tender in less than two weeks.

Publication

Depending on national regulations and the contract award procedure selected, a government agency that contracts out work must publish its intention to award the construction contract in advance. Advance publication of the information enables construction companies to request the tender documents and apply on time. › Fig. 17

Lead times for trade contractors

Not every construction job can be started immediately after the contract is signed. Construction companies must often carry out additional steps before they can perform the work on the construction site. Planners must also take this lead time for trade contractors into account, especially for jobs that require planning work by the contractor, as well as off-site prefabrication and extensive procurement of materials.

Planning material requirements

To ensure a secure financial framework when procuring materials, companies generally only place orders after signing a contract. For many construction tasks such as plastering and screed work—which use standardized and readily available building materials—companies can do so in the abovementioned two-week period between the signing of the construction contract and commencing construction.

But if companies require materials that cannot be purchased in standard forms at wholesale outlets, planners must consider and check in advance the time involved in planning material requirements.

Inspecting samples

If the owner wants to inspect samples (e.g. bricks, tiles, windows, colors and similar items) › Figs. 18 and 19 before orders are placed, planners must leave sufficient time for the following steps:

_ Procurement of samples
_ Inspection and approval of samples
_ Modification or procurement of additional samples (if necessary)
_ Material delivery periods

Prefabrication

In addition to the time needed to procure materials, some construction work requires planning by the construction company and involves prefabricating components before the work can be carried out on the construction site.

Depending on the construction job, the construction company may have to take on-site measurements in order to prefabricate and install components precisely. Measurements can only be taken once there has been sufficient progress in the construction process (e.g. completion of floors or openings in the building shell).

Based on these measurements, the construction company creates its own working drawings, which provide a foundation for prefabricating the necessary components. If stipulated in the contract, the architect approves technical aspects of the working drawings before components

\\ Example:
In large companies, the staff in charge of construction work may not be able to finalize a decision if contract volume exceeds a certain limit. A higher-level body such as the management board must first approve the contract award. This may entail a long period of time, depending on how often this body meets.

\\ Example:
If natural stone tiles are needed in specific sizes from foreign countries, they must first be ordered, manufactured and imported by ship. If particularly large quantities of a special material or component are needed, or if single parts must be produced, production may take some time due to the lack of supplies at wholesale markets.

Fig.18:
Inspecting a facade system

Fig.19:
Inspecting the design of the roof edge

are manufactured. In this case, planners must schedule inspection and approval times in addition to the periods needed to create and work on the working drawings. › Fig. 20

After the architect gives the green light, prefabrication can begin. Depending on the trade contractor and the component, it can take six to eight weeks or more before the component is ready for assembly. However, the actual on-site assembly process usually takes a relatively short amount of time.

Elements that typically require prefabrication are:

_ Facades, windows and doors
_ Glass roofs and skylights
_ Precast concrete units
_ Steel structures (e.g. loadbearing hall structures, stairs, railings)
_ Timber structures (e.g. roof trusses)
_ System elements (e.g. office partition walls)
_ Ventilation systems
_ Elevator systems
_ Built-in furniture and interior installations

Construction period

The construction period encompasses all the tasks covered by the contract work package. When structuring such tasks, planners must consider the dependencies between components and trade contractors, as well as task duration and the assignment of work to construction phases. › Chapter Creating

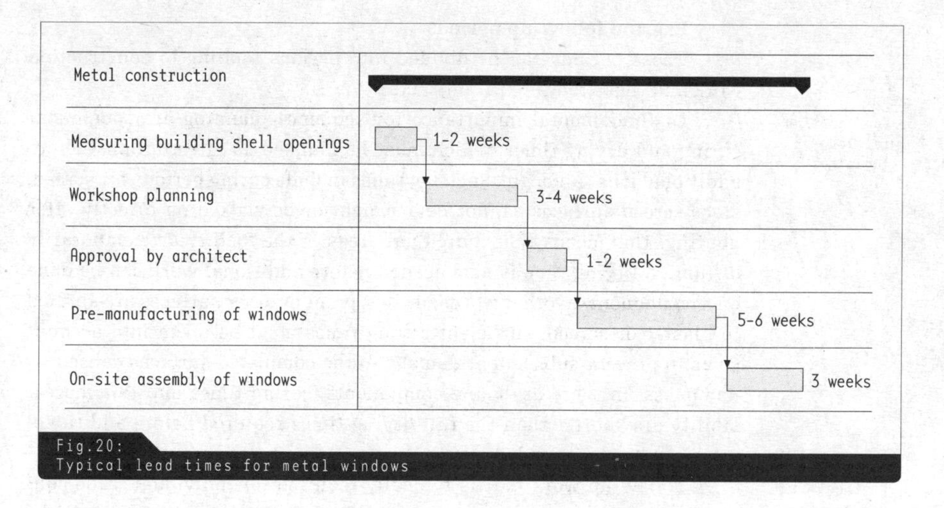

Fig.20:
Typical lead times for metal windows

a schedule, Planning task duration As a rule, the construction company should be allowed to schedule its work on a detailed level if there are no dependencies with other crews. In areas of the building where several trade contractors intermesh, the work should be organized in such great detail that each contractor can see its own working periods and interdependencies.

If the different jobs performed by one trade contractor are separated by longer periods and are not structurally linked, it can be effective to divide the skilled tasks into two contract work packages so that both an adequate planning lead time and self-contained, coherent construction contracts are ensured.

The tasks and dependencies that are typical of the planning and construction process are described in the chapter "Workflows in the planning and construction process."

\\Example:
Steel and metal construction often involves a number of jobs that extend through the entire construction process. These include the construction of steel structures, windows, exterior wall cladding, doors, railings and stairs. Since in most cases these jobs do not build upon one another and the companies usually specialize in certain fields, it is advisable to plan partially separate contract work packages.

Lag and follow-up periods

These periods can be divided into periods relating to construction work and those relating to contracts.

Construction-related lag times

Of fundamental importance for sequence planning are component curing and drying times, which must be planned as interruptions before additional jobs can begin. Such lag times include curing periods for screed, since screed surfaces cannot bear weight or be walked on directly after pouring. This means that individual areas of the facility are temporarily off limits. Drying time is also needed before additional work can be done on a component. In other words, if tiles, paint or other surfaces are applied to plaster or screed, sufficient drying times must be taken into account so as to prevent subsequent damage to the completed surfaces caused by dampness. In there are many components, curing times and thus accessibility are shorter than the full drying times required before additional work can be performed.

Contractual follow-up periods

Follow-up work that is typically required for individual trade contractors influences the contractually stipulated construction period of a contract work package. Examples of follow-up work are:

_ Building shell: closing wall openings after building services are installed, removing site installations after completion of the building (if contracted out with the building shell)
_ Windows and doors: assembly of windows and door handles before completion of the building
_ Plastering: plastering flush with doors, stair coverings, window sills
_ Building services: detailed installation work for switches, heating elements and sanitary fixtures; launching technical systems
_ Painting: additional coat of paint after tiling, detailed installation work and assembly of fixtures on finished walls and ceilings

Planners should enter such follow-up work into the schedule as separate tasks in order to eliminate the possibility of claims from construction companies that have exceeded contractually defined work times. Also of relevance are the guarantee periods that start with inspections. The sooner both the contractually stipulated work ends and the inspection takes place, the earlier the guarantee periods end that give owners the right to have damage repaired.

Follow-up periods in the field of planning primarily entail information and advisory services that become necessary when designs or conditions change during the construction process (e.g. unexpected discoveries in existing buildings or on the building lot).

PLANNING TASK DURATION

Once planners have recorded all the tasks of the various trade contractors and planning participants, they must now estimate how long tasks will last. Architects usually use empirical values from past projects or question trade associations and construction companies about typical task durations.

Another way to calculate durations is to use defined quantities and quantity-related time values. Here an important distinction must be made between unit production time and unit productivity rate.

Unit production time

The unit production time (UPT) indicates how many person hours are needed to produce a unit of work. It is calculated as follows:

Unit production time = required person hours / quantity unit (e.g. 0.8 h/m^2)

Unit productivity rate

The unit productivity rate (UPR) is the reciprocal of the unit production time. It indicates the quantity produced per time unit:

Unit productivity rate = executed quantity / time unit (e.g. 1.25 m^2/h)

In construction, unit productivity rates are used primarily for equipment (e.g. the performance of a power shovel expressed as m^3/h), while unit production times are applied to labor (e.g. the number of hours needed to make a cubic meter of masonry, expressed as h/m^3).

Determining quantity

Estimations of quantities are based on the quantity units of the underlying unit production times and productivity rates (m, m^2, m^3, or piece). If a productivity rate is expressed in terms of cubic meters of earth, excavations must also be calculated in cubic meters.

Since the quantity units are often the same as those in other stages of the planning process (costing, tendering, etc.), the quantities can be

\\Note:
Unit production time and unit productivity rate are always dependent on the type of construction company, its working method, and the workers involved. Furthermore, on-site work is often influenced by the specific conditions there. It is therefore never possible to calculate precise task durations in advance using these approximate values.

adopted directly from these other stages. If no quantities are available, they must be calculated from scratch. When choosing the degree of detail in quantity calculations, planners should keep in mind the imprecision of unit production times and unit productivity rates. A rough calculation is normally sufficient.

Establishing task duration

The total number of work hours necessary to perform a task—known as person hours (PH)—can be calculated on the basis of the required quantity and either the unit production time or unit productivity rate. If person hours are divided by both the number of workers (W) and the daily working time (DWT), the result is the probable task duration (D), expressed in workdays (WD):

$$D = \frac{PH(UPT*quantity)}{W*DWT} = [WD]$$

Number of workers

The daily working time is usually set by the regulations in collective wage agreements. Overtime must be allowed for in special circumstances, such as significant deadline pressure. All things considered, an optimal number of workers should be allotted so as to ensure an effective construction process. Some jobs such as window assembly require a minimum number of workers—otherwise they cannot be done properly or cannot be done at all. Nevertheless, worker numbers cannot be randomly increased since there is a chance that the workers will then no longer be effectively deployed. One example is screed work, where the number of workers depends heavily on the availability of equipment, the productivity of which can be only slightly increased using more personnel.

The specification of staffing capacity is merely an internal calculation method used to achieve a reasonable implementation duration. Normally planners leave it to the construction companies themselves to deploy an adequate number of workers for the available construction period. However, such calculations can help site managers determine

\\Tip:
Depending on construction conditions, published time values and real values may deviate by up to 50%. Smaller quantity differences in calculations can therefore be seen as negligible. If precise information is required, planners should compare several sources for time values. The appendix contains a summary of many typical unit production times as a basis for calculation.

\\Example:
If the unit production time is 0.8 h/m^2, the quantity 300 m^2, the number of workers 5, and daily work time 8 hours, task duration is:

$$D = \frac{0.8 \times 300}{5*8} = 6\ WD$$

whether a site is understaffed to the extent that problems will arise in meeting final deadlines. The reverse of the equation can be used to figure out the number of workers necessary to complete a job in the given timeframe.

$$W = \frac{PH(UPT*quantity)}{D*DWT}$$

Such calculations can also be used to evaluate contract awards with reference to the on-site performance capacity of construction companies.

It is an advantage for the scheduler if worker numbers are tailored to task durations in the given construction phase. If several trade contractors work one after the other in one construction phase and then gradually switch to the next phase, identical task durations will ensure that work is constantly being performed in each phase and that crews do not have to wait for each other. › Fig. 21

The results of duration planning

It is not necessary to calculate task durations precisely for every single schedule and every single task. Often all that is needed is an estimate based on empirical figures. The main reason is that every construction project is subject to small modifications of task durations, and in most cases these have only a limited effect on the completion deadline for the entire building. What are much more important for meeting overall deadlines are task sequences, which are described in the next chapter. Here errors can cause structural displacements and have far-reaching consequences for the project.

Nevertheless, task durations must not be overlooked since they provide a basis for the implementation deadlines that are agreed upon in the contracts with construction companies. It is therefore crucial to use realistic, viable task durations in order to facilitate smooth execution of the various construction contracts.

\\Example:
If a total of 240 person hours is required to do a job and the given timeframe is limited to five working days, the number of workers required is calculated by dividing 240 PH by 5 x 8 (PH/D x DWT). The result is 6 workers.

\\Tip:
When determining task duration, planners should not only take a mathematical approach, but also consider other factors that keep the construction work from proceeding at the same rate as during the rest of the year (holidays, typical vacation periods, frost periods, etc.). This is typical of the time around Christmas and New Year's Eve, despite the fact that a sufficient number of working days are available.

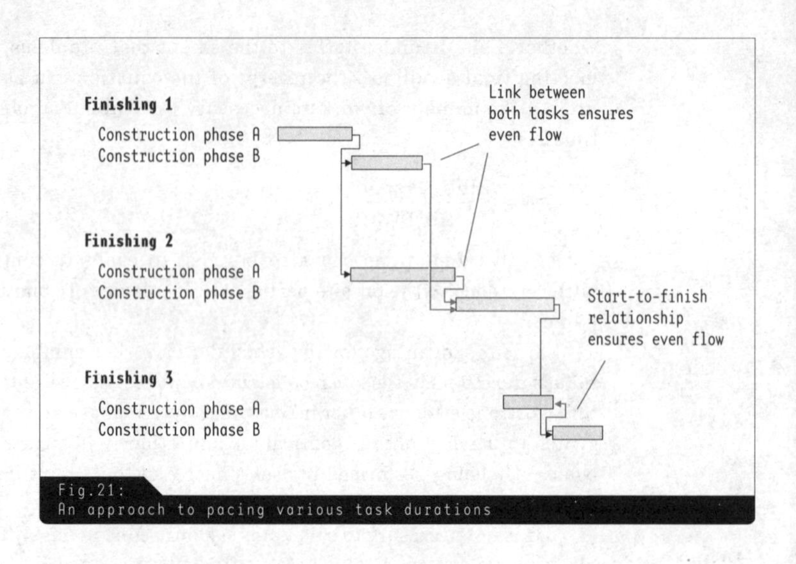

Fig.21:
An approach to pacing various task durations

The results of duration planning

It is generally impossible to calculate planning services using unit production times, since intellectual and creative work cannot be grasped in terms of hours per quantity unit. The durations of planning tasks are usually determined in conversations with planners and experts when they are hired and as the project progresses. This approach optimally exploits the planners' personal experience and available time. By illustrating the effects of possible delays in planning stages, it also makes the participating parties conscious of their own contribution to completing the project on time.

WORKFLOWS IN THE PLANNING AND CONSTRUCTION PROCESS

The following chapter describes typical tasks performed by the participants in the planning and construction processes and discusses the dependencies of these tasks on one another. Its goal is to use this background information to examine projects, identify relevant tasks and depict them in a real-world schedule.

PLANNING PARTICIPANTS

The various parties that need to be coordinated in the planning phase can be roughly categorized as follows: › Fig. 22

Owners and affiliated parties

To begin with, the owner or client must be mentioned, as the initiator of the building project. The owner may be a single person or a complex web of people and institutions. These different constellations can result in significantly different perspectives on the part of owners, project developers, financial backers (banks), and subsequent users. For example, if the owner is a company or a public institution, the project supervisor must answer to committees and departments that have both influence and decision-making power and that must be integrated into the owner's decision-making processes.

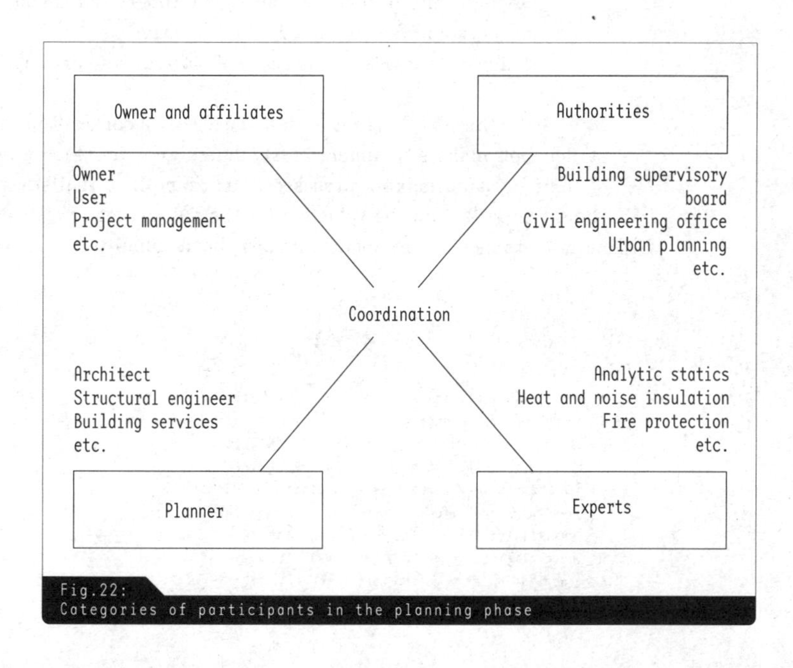

Fig. 22: Categories of participants in the planning phase

In efforts to coordinate the people involved in the owner's company or organization, it is important for planners to understand the decision paths and to estimate the time involved so that they can make punctual preparations for decisions.

Authorities

Further, every newly erected building requires interaction with the authorities that decide on the legality of the project, grant permits under public law, and monitor the project using tests. The extent to which authorities intervene in the process depends largely on the function of the planned facility, the building type, the legal regulations, as well as local conditions. In addition to building supervisory boards, the project may involve the following authorities:

_ Civil engineering offices (connecting the property to public infrastructure)
_ Urban planning authorities (analyzing the urban planning context)
_ Environmental authorities (the environmental effects of the project)
_ Occupational safety and health authorities (worker safety on the construction site and later in the completed building)
_ Historical preservation offices (for historical buildings)
_ Land surveying offices (maps, site plans)
_ Registry and property authorities (property management, encumbrances and restrictions on the property)
_ Trade authorities (in the event of subsequent commercial use)

Since these authorities normally exercise a control function or serve as decision-makers, planners must understand the steps and durations of their decision-making processes. For example, a realistic time period should be built into the schedule for the authorities to award a building permit—to begin after documents have been submitted.

\\Example:
If, for the additional planning process, architects require an decision from the owner on floor coverings or other surfaces, they should provide the owner at an early stage with samples of viable alternatives with corresponding advantages and disadvantages (costs, lifecycle, sensitivity, etc.). Owners may have to discuss these alternatives with other persons in their organization or with subsequent tenants.

Planners

The planning staff consists of various project planners and planning specialists. The project planner (usually the architect) brings them together and resolves any possible conflicts between the different requirements. The three most important integrated planning areas that extend through the entire planning process are the architecture, structural engineering and building services. However, in practice, a large number of planning specialists may be involved:

- Structural engineering
- Interior architecture
- Electrical engineering
- Drinking and waste water engineering
- Ventilation systems
- Fire protection engineering
- Data technology
- Elevator engineering
- Kitchen planning
- Facade engineering
- Landscape and open space planning
- Lighting systems
- Facility management

Experts

Along with the planning participants, experts are required to assess and submit reports on the different specialized tasks. At the very least, this group of experts includes specialists who assess and test heat insulation, noise protection, fire protection and statics.

The experts' assessments must be taken into account in the schedule, primarily in connection with the submission of results. For example, specific expert opinions must be submitted for the construction permit and the start of construction. Experts must therefore be hired and given the work documents with an appropriate lead time.

\\Note:
A detailed description of the participants in the construction process and of the planning processes that build upon one another can be found in: Hartmut Klein, *Basics Project Planning*, Birkhäuser Verlag, Basel 2007.

COORDINATING PLANNING

The intensity of coordination during the planning phase is very much dependent on the size and complexity of the structure in question and on the scheduling constraints imposed by the owner. In the case of a residential building, which is largely planned exclusively by the architect, only a few deadlines are usually relevant, such as the application and award of the building permit and the start of construction. In larger structures, such as laboratories and specialized production facilities, input from a range of specialists is often required.

The way in which the planning phase is divided up is only partly dependent on the sequence of architectural planning, since other participants will have structured their particular areas differently. The best way to approach the organization of planning is by making an initial link between the three most important planning areas—architecture, structural engineering and building services—since, as a rule, planners in these areas have a constant integrated influence on the planning process as a whole. Planning should take into account the fact that planning specialists require an appropriately advanced state of overall planning in order to be able to make their particular contributions. This networked approach is typified by the following sequence:

1. Advance development of an appropriate foundation for planning specialists by the project planner
2. Forwarded to planning specialists
3. Worked on by planning specialists
4. Results returned by planning specialists
5. Integrated by project planner and coordinated with results of planning by other participants
6. Mutual coordination of specialist planning and (if necessary) further revision

The schedule should allocate enough time after the submission of responses by planning specialists to resolve inconsistencies between specialist planning and project planning or other areas of specialist planning. › Fig. 23

\\Tip:
Typical information exchanges between the architect, structural engineer and building services engineer relating to different project phases can be found in the appendix. However, the details of such exchanges differ according to the specific project and depend on the function and design of the building and the people involved.

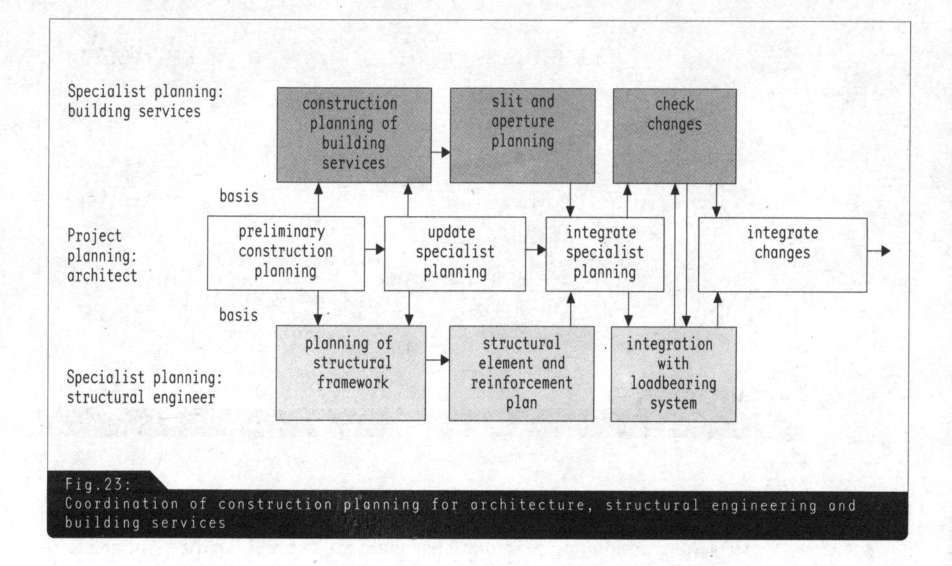

Fig.23:
Coordination of construction planning for architecture, structural engineering and building services

Integration of other participants

Integrating other participants is generally simpler, since their work is not as closely linked with that of the planners referred to above. It is often sufficient simply to include a period in the schedule in which planning specialists will be given time to react to particular results of the overall planning process. This will clarify when the various participants need to be given the information they require for planning their specific input into the project, and when the results will be integrated into the overall planning process.

Planning milestones

The most important aspect of the scheduling process is the identification of milestones that can be passed on to the various participants (building owners, planners, experts) in order to provide them with a basis for their respective input into the project. › Chapter Creating a schedule, Schedule elements Established and agreed deadlines ensure that all parties adopt the discipline of a fixed timeframe and help avoid the delays in the planning process that can occur when participants are included in the process too late.

COORDINATING CONSTRUCTION PREPARATION

When it comes to awarding contracts, planning deadlines must be particularly well thought out. Construction preparation is a time-consuming process that can easily take several months. If it is clear when a building company has to begin work on the construction site, this deadline can be used as a basis for scheduling the different steps in the preparation process. › Chapter Creating a schedule, Planning task sequence If owners are public bodies, legally prescribed deadlines mean that planning can include only limited provisions for delays, which will thus directly affect the start of construction.

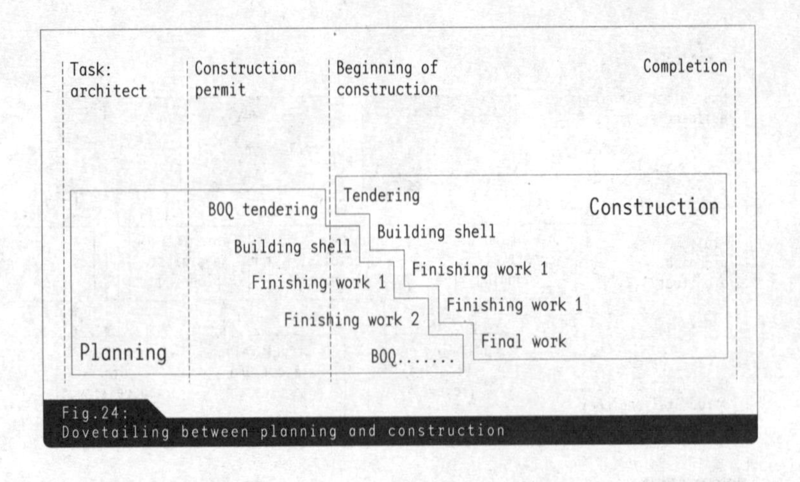

Fig.24:
Dovetailing between planning and construction

Since a range of companies are usually involved in a building project, the construction preparation phase must make provision for a range of different deadlines. It is therefore wise to organize the respective steps in the preparation process to ensure that all preliminary work, such as securing owner decisions, construction planning, and organizing and announcing tenders and contract awards, can be punctually initiated in relation to each trade contractor involved.

Dependencies generated by construction workflows also give rise to planning requirements.

Planning during construction

Many rough schedules present planning and construction as separate phases, but in reality, construction planning and construction itself largely proceed in parallel. This is because deadlines are often very tight, and planning does not necessarily have to be completed when building commences. Although it is important that documentation is available in good time before the relevant construction work commences, many types

\\Example:
When the building shell is planned, consideration needs to be given to the subsequent effects of facade connections, ceiling and stairwell heights, as well as the surface treatments of concrete walls and their associated requirements. Drainage underneath the bottom slab also needs to be clarified at a very early stage.

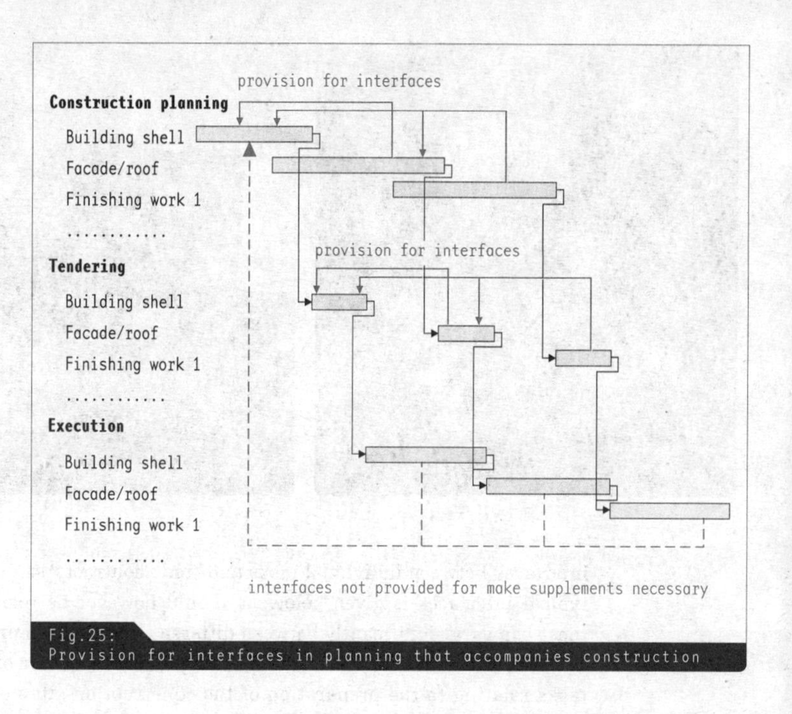

Fig.25:
Provision for interfaces in planning that accompanies construction

of work (such as painting and laying flooring) tend to commence once the building process is quite advanced, which means that relevant planning and contractual documents can be completed after construction has begun. In such cases, owe refer to planning that accompanies the construction process. › Fig. 24

In planning that accompanies construction there is always a risk that specific planning details that are developed later in the process will have an influence on aspects of construction that have already been completed. Many building elements interface with other parts of the building in terms of statics, building services, structure and aesthetics.

Since individual building elements such as windows, doors and dry construction work are increasingly integrated with other work sections, planning for such elements needs to be sufficiently advanced in order to avoid the need for subsequent modifications. › Fig. 25

PREPARING THE CONSTRUCTION PROCESS

During the construction process the building planner or site manager needs to coordinate all companies operating under a separate contract with the building owner. The goal is to ensure that work proceeds in an integrated and trouble-free manner. A central aspect of this coordination involves the

Fig.26:
Manual demolition on projects involving existing buildings

Fig.27:
Mechanical demolition of entire buildings

interfaces between individual tasks and trade contractors. A description of typical interfaces is given below; it should however be remembered that these can vary significantly between different construction projects.

Prior to the erection of the building shell, a number of preliminary tasks relating to the preparation of the construction site need to be considered. First, the site needs to be in a condition that actually allows construction work. Vegetation must to be removed and the ground reinforced, pre-existing piping and drainage systems must be located and protected, and pre-existing constructions (walls, fences, etc.) need to be demolished.

Preparing the construction site

The required preparatory measures include setting up the construction site. This process involves installing the construction trailers from which construction is supervised, connecting utilities, and putting up fences to prevent unlawful entry. Further work may be required to improve access (e.g. access roads) and to shield the site from the external environment (e.g. blinds and noise protection).

Demolition measures

For new buildings, preparatory measures can often be completed within a few days or weeks. However, in projects involving existing building stock, the need for extensive demolition work may require scheduling a significantly longer preparatory phase. At the same time, the unpredictability of the demolition process often makes it extremely difficult to estimate task durations. The choice of the demolition method decisively influences task durations, since mechanical demolition with heavy equipment cannot be compared with manual demolition using light machines. › Figs. 26 and 27 Routes for waste transport within the building and access to dumps also need to be taken into consideration. Because demolition measures precede construction, delays in the demolition phase directly affect all subsequent work.

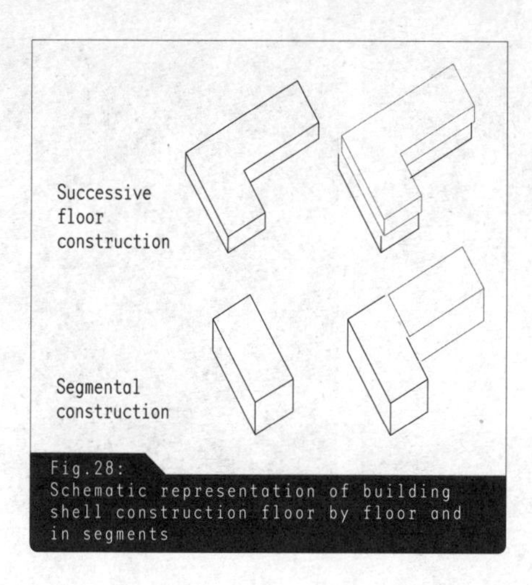

Fig.28:
Schematic representation of building shell construction floor by floor and in segments

BUILDING SHELL

Constructing the building shell entails a range of tasks, all of which contribute to creating the building's basic skeleton. In the case of concrete construction these tasks include:

- Excavation
- Masonry
- Pouring concrete
- Putting up scaffolding
- Sealing the building to protect against rising damp and groundwater
- Separate roof construction (if required)

Depending on the type of construction, structural steelwork and timbering may also be needed. As a rule, building shell construction is organized by the construction contractor. Of primary concern for architects are the interfaces with subsequent work sections that commence after the shell has been completed.

The sequence of tasks involved in the construction of the building shell is usually highly structured and easily comprehensible. Once the foundation and drainage systems have been completed, the floors are added in succession. However, where large floor areas are required, building shell construction may involve working in vertical segments as well as successive floor construction. › Fig. 28

Pre-manufacturing in the building shell phase

If a building is made up of elements such as prefabricated concrete units and steel or wooden structures, these components are usually pre-manufactured off site, delivered ready for assembly and quickly mounted

Fig.29:
Construction of a concrete wall using prefabricated elements

Fig.30:
Construction of bottom slab and concrete supports cast on site

on site. › Fig. 29 and Chapter Creating a schedule, Planning task sequence Apart from the different floors, this usually also applies to roof structures made of wood or steel. These structures may have to be contracted out separately from the building shell, and this needs to be taken into account in the construction schedule.

BUILDING ENVELOPE

Envelope sealed

Directly after the completion of either the building shell (including roof structure) or the individual segments of the shell, the building needs to be sealed off from its surrounding environment. Sealing the building envelope is a basic precondition for all subsequent finishing work. For this reason the "envelope sealed" stage should be reached as soon as possible after completion of the building shell. The relevant functional requirements are:

_ Rainproofing (protection for finishing elements, drying out the building shell)
_ Windproofing (above all in winter, to retain heat in the building)
_ Security (protection against theft of finishing elements)
_ Heatability (only necessary during winter months)

Windows and doors

The most important precondition for reaching the "envelope sealed" milestone is the sealing of openings and roofs. This is achieved either by installing windows and doors immediately or by blocking off openings using temporary seals such as guard doors. Depending on the type of construction, builders can add insulation and the facade of the closed external wall immediately after the "envelope sealed" stage. Where additional, thick

Fig.31: Assembly of continuous windows

Fig.32: Assembly in attic area

outer-wall features are added, it may be necessary to modify or shorten scaffolding.

Sealing roofs and drainage

The covering of steep roofs and the sealing of flat roofs must be completed in order to seal the building envelope. This also applies to skylights and roof-light domes and includes the completion of plumbing and roof-drainage work. For drainage embedded within a flat roof, builders must also ensure that once the envelope is sealed, water is actually drained out of the building.

Lightning protection

Before scaffolding for facade and roofing work is removed, lightning protection work also needs to be carried out and lightning conductors must be earthed.

FINISHING WORK

Coordinating finishing work is the most demanding segment of the construction supervision process. Precise scheduling is essential since tasks are closely linked and the different types of work being simultaneously carried out are often not restricted to one or a few companies—in contrast to the building shell and envelope.

Most contractors will not be able to coordinate their own work with that of other contractors because of the complexity involved. For this reason, the schedule must define the relevant dependencies in detail.

Plastering work is generally carried out soon after the building envelope has been sealed. Since in most cases wall installations need to be concealed behind plaster surfaces, they must be in place before plastering begins. In this context care should be taken not to overlook features such as drive mechanisms for doors, fire protection installations and emergency lighting. In industrial buildings, cables are usually laid on the outside of walls, which means that plastering can be done before building services are installed.

Plastering

Door frames are a typical interface in this context, since the type of door frame chosen determines whether it is mounted before or after plastering. For instance, steel corner frames should be mounted prior to plastering, because embedding frames and intrados later on usually generates additional costs. Dual section closed frames should only be installed late in the construction process to protect against damage. Interfaces that can influence the sequences of all types of work take place wherever building elements intersect with one another, as is the case with windows, window sills, stairway access points and stairway railings. › Figs. 33 and 37, p. 56

In order to ensure that surfaces are ready for further work (e.g. painting), the schedule needs to include a drying period appropriate to the type and thickness of the plaster used.

It is also helpful to include a post-plastering section in the schedule to ensure that surfaces damaged by other work can be repaired at a later date.

Screed work

Schedules usually make special provision for screed work, because no other work can be carried out while screed is being laid or is curing. As a consequence, the schedule must not only cover screed application but also include a curing period appropriate to the type of material used. Cement screed has a curing time of three to ten days (depending on the aggregate, weather and thickness), and it is significantly more economical than, for example, poured asphalt screed. However, the latter is ready for use and further work after only two days.

Apart from curing time and readiness for use, another important aspect of screed work is drying time. Flooring can commence only after screed has dried sufficiently and moisture content is low. The drying time is primarily dependent on the type and thickness of the screed and on environmental factors such as temperature and humidity. In many cases flooring is only installed late in the construction process to avoid such problems. In cases where deadlines are tight, the drying time can be reduced by using additives or drying equipment, although such measures generate additional costs.

When scheduling screed work, planners should also consider installations laid under screed, such as floor heating, heating distributors, floor power outlets and electrical conduits. › Fig. 34 Due to the inaccessibility of building sections containing freshly applied screed, interconnected workflows need to be checked in terms of possible system bottlenecks, walkways (and escape routes), transport routes for materials, as well as intersecting installation areas (such as electrical conduits).

Drywall

The sequence in which drywall construction and screed work are carried out also depends on whether more priority is given to noise insulation

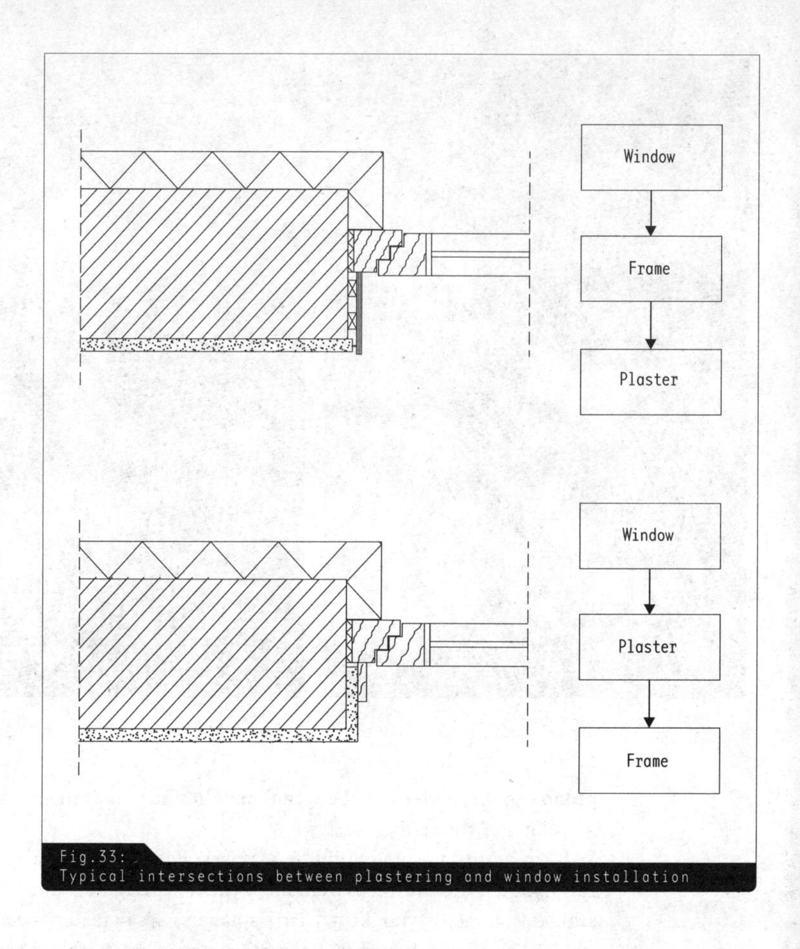

Fig.33:
Typical intersections between plastering and window installation

(walls attached to the unfinished floor) or flexibility (walls attached to the screed). Where a large proportion of the walls and ceilings in a building are constructed using the drywall method, the coordination of this work with many other work sections is one of the most important tasks in the scheduling process.

Due to these dependencies, plasterboard walls are usually constructed in two steps. First, the substructure is erected and sealed on one side. This is followed by all installation work related to building services (electricity, sanitation, heating, ventilation). The drywall contractor only seals the second side of the wall once this work has been completed.

The construction of suspended ceilings also involves close integration of installation work and drywalling. All raw installations must be completed before the ceiling is mounted, and planners need to consider possible

Fig.34: Installations prior to application of elevated screed

Fig.35: Rough installations prior to closing a suspended ceiling

geometric dependencies between installations and the suspension and screening of the ceiling. › Fig. 35

In addition, some building-services elements require specific preparation of drywall surfaces and subsequent sealing once they have been installed (e.g. recessed lighting, access panels, fire detector covers). › Fig. 36

Doors and partition walls

With plasterboard walls, frames are often installed while the wall is being constructed, because they must be screwed to and aligned with lateral profiles. In the case of solid walls, the frame is usually fitted before or after plastering in the form of a corner, profile or dual section closed frame. › Fig. 37, p. 56

In addition to affecting the way doors are mounted (in the building shell and the drywalling work), the type of frame or door construction is a significant factor in determining the point at which mounting should take place. For normal doors, frames are installed before or after plastering or during drywalling depending on the particular situation, and the door leaf is mounted as late as possible to avoid possible damage. Metal frame doors and system elements as well as standardized steel doors and panels are often supplied and installed as complete units, including the frame and door leaf.

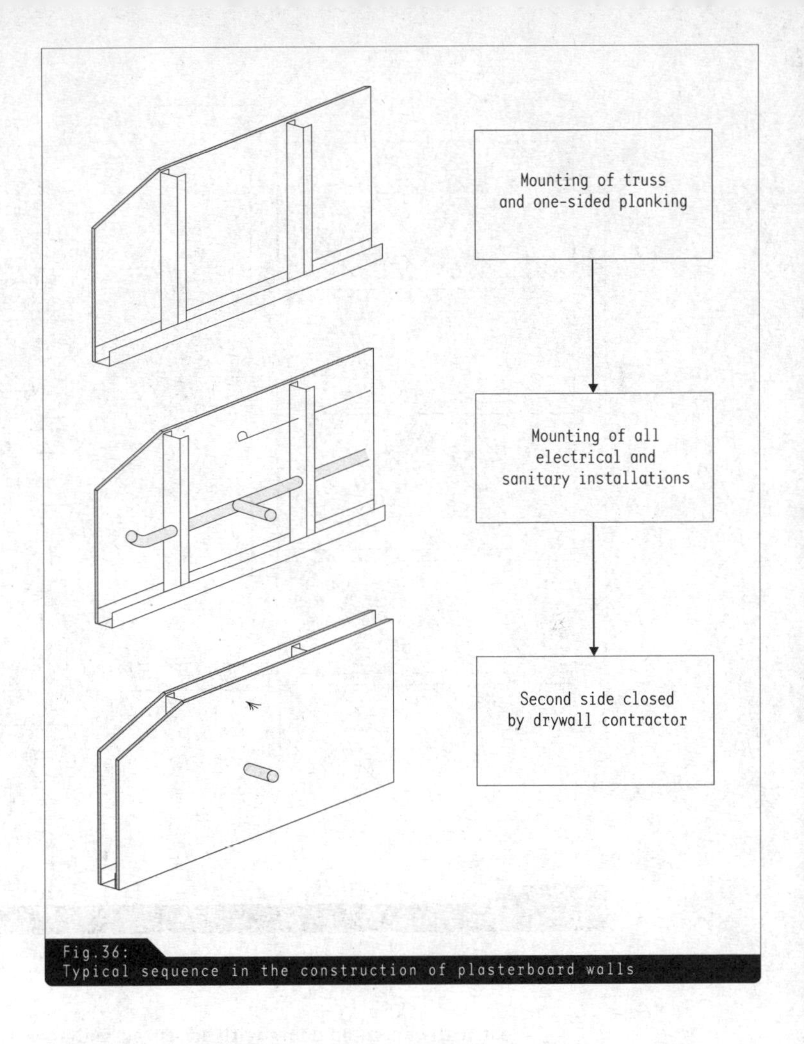

Fig. 36:
Typical sequence in the construction of plasterboard walls

In many cases, door details have a decisive effect on the time period allocated for installation:

_ Door frame installation with or without floor recess (dependent on screed)
_ Door structures with or without surrounding frames (dependent on screed)
_ Frame geometry: frame over plaster or plaster applied to frame (dependent on plastering) › Fig. 37, p. 56
_ Permit requirements application of plaster to fire doors

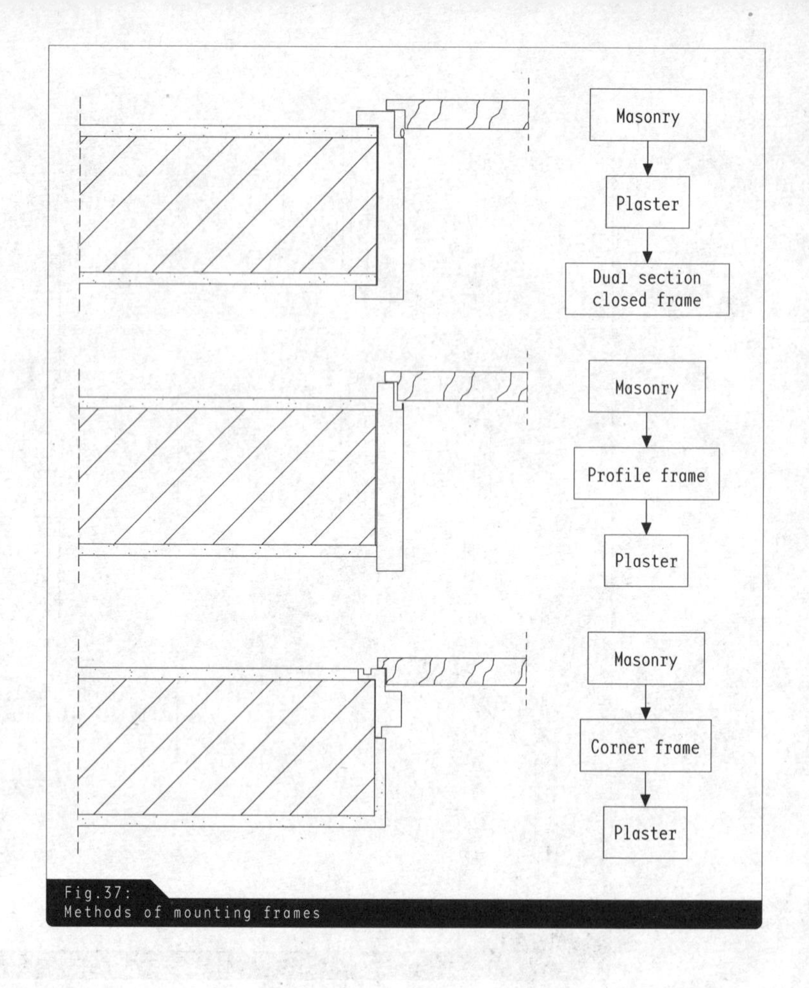

Fig.37:
Methods of mounting frames

- Electrically operated doors with access surveillance, escape route functions, accessibility for disabled people, automatic door openers (dependent on electrical systems and fire alarm installation)

In principle, sensitive installation elements should be planned as late as possible in order to avoid damage to finished surfaces.

In addition, depending on the type of door, delivery times often must be taken into consideration. Standardized steel doors and frames can be delivered quickly and installed as complete units. Special structures such as fire doors and complete metal-frame door units are manufactured to order, and delivery times can easily be 6 to 8 weeks or more.

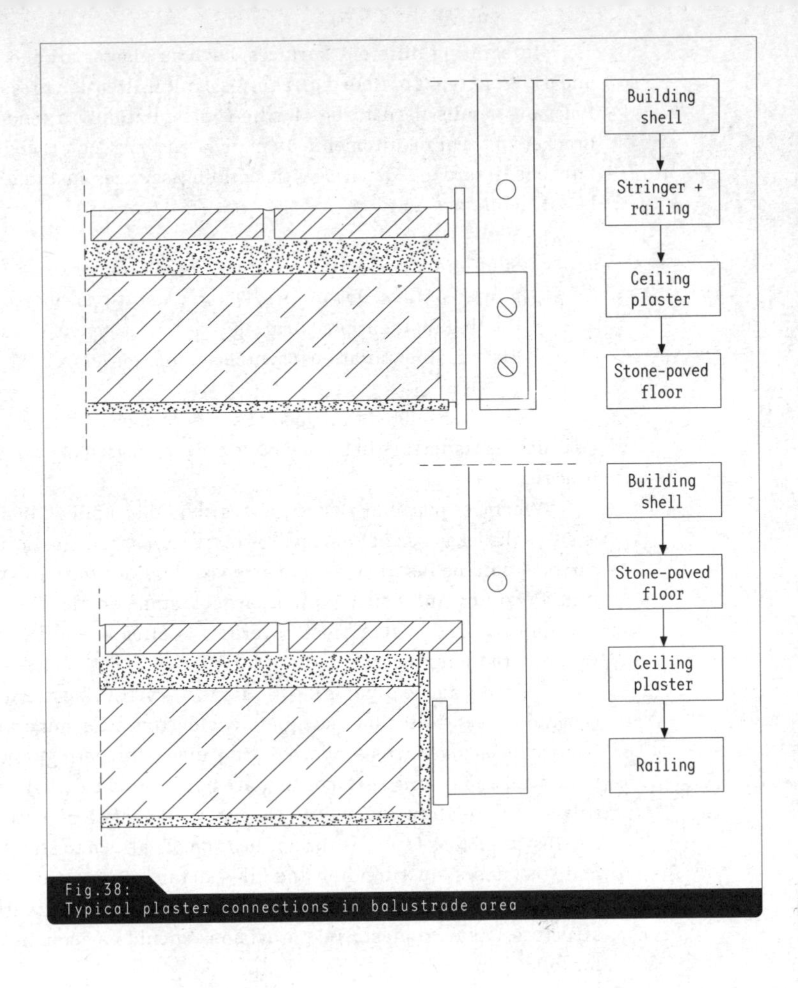

Fig.38:
Typical plaster connections in balustrade area

Tiling, parquet laying, stonework and flooring

A fundamental requirement for laying tile and stone surfaces is the completion of the underlying foundation. However, a distinction must be made here between using a thin bed of plaster on a level foundation, and a thick bed on the building shell surface.

Different surfaces, such as screed, building shell surfaces, masking, plaster, drywall etc., can serve as the foundation for all coverings and coatings. For stairs, the sequence in which a covering is applied will also depend on the way the stair railing is attached, the possible addition of a stringer, and in some cases the presence of scaffolding in the stairwell. **› Fig. 38**

Integrating different surfaces, such as plaster around door frames and different types of flooring, requires careful attention to task sequences. Relevant details need to be clarified in invitations to tender, including bracket and seal requirements. In many areas, particularly bathrooms and kitchens, interdependencies with building services installations need to be taken into account:

_ Sanitation installations: rough installations such as toilet cores, ground outlets, downpipes, water connections, access openings
_ Heating installations: heating pipes, heating elements
_ Electrical installations: switches, floor outlets, etc.

It should be noted that special surfaces such as flooring in elevators and tiled backsplashes in kitchenettes are often overlooked in the planning process.

Wherever possible, floor surfaces should be applied in an order that ensures the least risk of damage. For instance, carpet, plastic and linoleum floors should be installed as late as possible, since they are more susceptible to soiling and damage than parquet, stone or tile. They can also be laid quickly. Laying such floor coverings is often one of the last tasks in the construction process.

Painting and wallpapering

Painting and wallpapering require dry level surfaces. For this reason, schedules need to include adequate drying times for mineral-based surfaces such as plaster. As a rule, painting and wallpapering work is relevant to all wall and ceiling surfaces that are not covered by other surfaces such as tiles or prefabricated ceiling elements. Planners also need to consider smaller-scale tasks such as the application of varnish to stair rails, frames and steel doors, dust-binding and oil-resistant coatings to elevator shafts prior to elevator installation, and fire- and rust-resistant coatings to steel structures. As with plastering, provisions should be made in the schedule for follow-up work.

\\Example:
Heating elements represent a typical interface problem. Planners need to be aware that heating is required on the construction site over the winter months in order to ensure that subsurfaces and paintwork dry properly. However, in some cases heating elements must be removed again later on, to allow the wall areas behind them to be painted.

BUILDING SERVICES

Building services include all installation work involving heating, water supplies and drainage, sanitation units, ventilation, electrical installations, data technology, fire prevention installations, elevators and other building-specific installations. Coordination of building services and their integration with interior finishing work are generally based on collaboration between building and building services engineers. In this context, it is important that building planners understand where the interfaces are located between different building services contractors and integrate these interfaces into workflows. › Fig. 39

Heating installation

Heating installation involves a range of different construction elements. The order in which these elements are installed during a particular project must be determined on the basis of the different systems and distribution networks. Typical elements are:

_ Energy supply (gas pipes, solar collectors, pipes, etc.)
_ Storage facilities (e.g. hot water and oil tanks)
_ Heat stations and heat generation
_ General distribution and sub-distribution within the building (duct installations)
_ Distribution per heating unit (heating unit connection)
_ Heating unit installation

Planners need to coordinate the installation of the heating system with relevant interior finishing work on a range of surfaces. If the heating system is to be used to heat the construction site during the winter months, parts of the system need to be installed in advance and then temporarily removed to permit work on surrounding surfaces (plastering and painting).

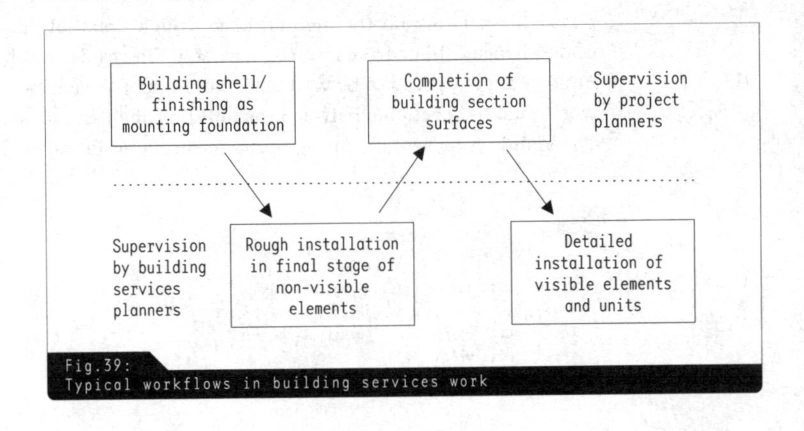

Fig.39: Typical workflows in building services work

Sanitary installations

Like heating systems, sanitation systems require closed and sometimes pressurized circuits and networks. For this reason, the workflows involved are similar to those in the installation of heating systems. Apart from the connection of the building to the local water provider, planners dealing with the installation of the drinking water supply need to understand how it is distributed throughout the building via feed pipes, ascending pipes in installation shafts or wall slits, and connections to individual use points in bathrooms, kitchens and similar rooms. Analogous planning must be performed for sewage disposal.

The rough installation phase is followed by the detailed installation of sanitation facilities (sinks, toilets, faucets, etc.). This work is often scheduled well after the completion of tiling and painting work › Fig. 41 in order to avoid the risk of theft or damage. Elements such as bathtubs and shower trays that are to be tiled on the outside need to be installed before tiling work commences.

Typical interfaces requiring coordination between work sections:

_ Piping underneath the bottom slab (often installed during building shell construction)
_ Wall and ceiling openings (which in some cases need to be closed after installation during building shell construction)
_ Ventilation ducts running through the roof (connection of sewage disposal pipes to fans in the roof area)
_ Water and sewage connections to the building (requiring coordination with public providers)
_ Installation of pumping systems below the backwater level
_ Heating water (installed centrally with a parallel piping configuration or decentrally at the point of use)

Electrical installations

Electrical installation work is also divided into rough and detailed phases. Depending on the method used, rough installations can be concealed beneath plaster or left exposed. If, as in residential buildings, cabling is not supposed to be visible, completion of the entire electrical network must be scheduled between the building shell and plastering phases. The visible (exposed) mounting often encountered in industrial buildings

\\Note:
Further information on drinking water and sewage system components can be found in: Doris Haas-Arndt, *Basics Water Cycles*, Birkhäuser Verlag, Basel 2008.

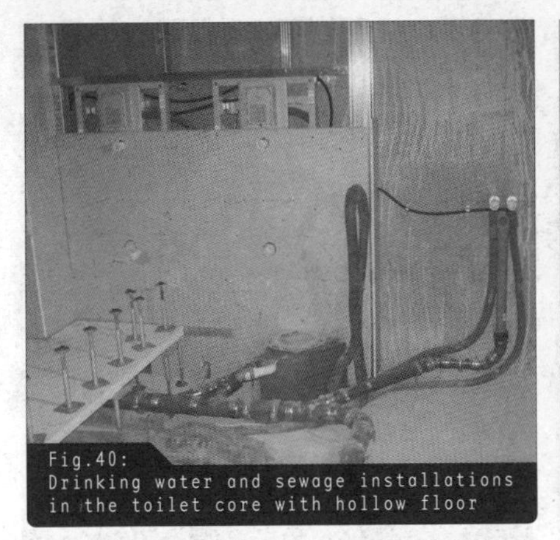
Fig. 40: Drinking water and sewage installations in the toilet core with hollow floor

Fig. 41: Tiled surface ready for detailed installation of sanitation facilities

is usually carried out only after all wall surfaces have been finished. Exposed concrete walls are a special case, in that ductwork must already be laid in the wall when it is reinforced in order to allow for later electrical installations. › Fig. 42 Typical stages in electrical work are:

_ Building connection and main fuses (coordination with electricity provider)
_ Earth connection (coordination with building shell work)
_ Battery and transformer installations (if required)
_ Distribution within building and sub-distribution to individual points
_ Detailed installation of lights, outputs, switches etc.

Due to the increasing integration of structural elements with electrical systems, scheduling electrical work is becoming an increasingly complex task. In order to avoid having to open finished surfaces for additional installation, planners need to systematically check construction elements detailed in the building design for possible interfaces with the electrical system. Typical examples are:

_ Specific connections for stoves, tankless water heaters, heating systems and particular structural elements
_ External lighting
_ Emergency lighting
_ Fire alarm units
_ Ventilation units
_ SHEV units (window, roof opening, smoke extraction)

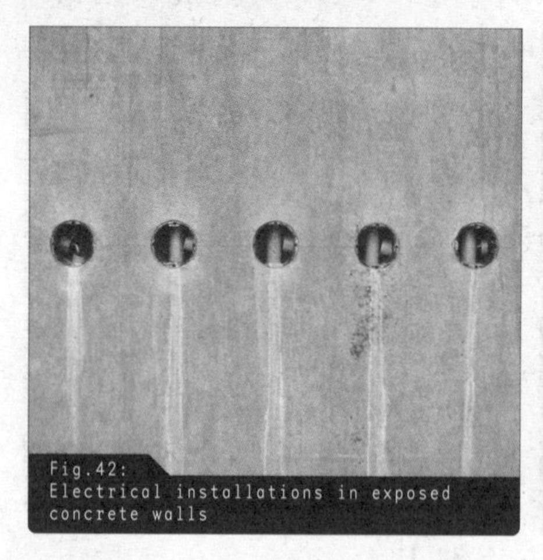
Fig.42:
Electrical installations in exposed concrete walls

Fig.43:
Example of distribution cables in floor

_ Disabled access (switches, automatic overhead door closers) and access monitors
_ Alarm system components (outer doors and emergency exits, window burglar alarms, surveillance cameras, etc.)
_ Facade control (electrical ventilation, sun and glare protection, rain and wind detectors, overhead lights, etc.)

Data technology

Data technology is a special aspect of electrical installation and is particularly complex in buildings used for administration and communications. Data technology refers to all forms of telecommunications and media technology such as telephone connections, television technology, computer networks, server rooms, etc.

In order to provide flexible access to every workstation in administrative buildings, server rooms and wiring closets are often installed either centrally or on each floor. Planners need to take into account dependencies with installation channels in floors, ceilings and walls.

Ventilation systems

Planning the installation of ventilation or air-conditioning systems must also cover the installation of extensive ductwork for incoming and outgoing air. As a rule, ventilation ducts, whether exposed or concealed, are laid in shafts, floor structures and ceiling areas, and planners must ensure that such installations are properly integrated with relevant structures and surfaces. It is critical in this context to take into account aspects such as supply conduits (cooling pipes, electrical cables, inlet vents), penetrations from the outer surface (caulking by roofer or facade construction contractors) and fire compartments (mortaring by the building shell contractor or fire-retardant sealing by the drywall contractor).

Along with the individual interfaces between the distribution system and the ventilation plant, the schedule for later work in the construction process needs to accommodate the detailed installation of features such as exhaust outlets, grating, covers and screens. Such installation is generally carried out after surfaces have been finished.

An important consideration when scheduling the installation of large ventilation and air-conditioning systems is lead time for prefabrication. Apart from a few standardized duct cross-sections, the dimensions of ducts, intersections and units must usually be calculated on site during building shell construction and represented in an independent working drawing. The components are only manufactured once this drawing has been approved, and the entire process can take several weeks. It is therefore important that the ventilation system contract be awarded to a specialist company at an early point so that the work on the building site can be carried out punctually.

Transportation technology

Conveyor technology such as elevators and escalators usually requires a large number of electrical connections. Planners also need to take into account interfaces with floor coverings (inner covering of an elevator car, connection to door sills) and walls (elevator door embrasure).

In most cases, the anchorage points for a planned elevator must be defined by anchorage channels already installed within the shaft during the building shell construction phase. This enables planners to select the elevator construction contractor or manufacturing system at the earliest possible stage. Once the building shell has been completed, the shaft is measured precisely, and a working drawing for the elevator is made. Following the pre-manufacturing phase, installation often proceeds in several steps. First the load-carrying system is installed in the shaft, then the elevator car is mounted, and finally the electronic control system is put in and connected with the electrical system.

Scheduling also needs to consider whether the elevator will be used to transport materials during subsequent construction. However, such use of elevators is usually not advisable because damage to the interior of the elevator car is inevitable. As a result planners often make a conscious decision to complete all work on elevators in a late phase of construction.

FINAL WORK

Final contract awards prior to completion

Apart from the follow-up work by individual participants already referred to (painting, detailed installation etc.) there are also entire sets of tasks that are concluded only at the end of the construction process. These can include:

_ Final cleaning following the completion of all work and prior to handover to users

_ Locking systems (delivery and installation of the final locking and access systems for later users)
_ Completion of outside space (access routes, garden and lawn design, parking areas, signage, lighting and other outdoor installations)

Inspections stipulated by law

In principle it is wise to schedule in a period at the end of a project for the correction of defects and for formal inspections, since these take time and inspection should be completed before the occupation date.

Inspections include both contractually stipulated inspections and inspections as prescribed under public law. In the latter case the building supervisory board checks that the completed building observes construction regulations, and approves the building for use. Such inspections also cover technical installations such as fire prevention facilities and heating and air-conditioning systems, which in some cases have to be checked by outside experts.

WORKING WITH A SCHEDULE

Even though a schedule may present a detailed and coherent arrangement of the tasks making up a construction project, it is not a static framework that, once formulated, will necessarily remain unchanged until the completion of work. The construction process continually gives rise to particular situations that make it necessary to adapt the deadline structure. A construction schedule should therefore be seen as a tool that accompanies the entire construction process.

UPDATING AND ADJUSTING A SCHEDULE

Conception versus reality

Real conditions on a construction site often look different from the situation envisaged in the schedule. There are many reasons for disruptions and necessary structural changes. › Chapter Working with a schedule, Disruptions to the construction process Schedules, which are usually printed on paper and displayed on the construction site, age quickly, resulting in work that no longer reflects their constraints. This makes updating the schedule essential. Ideally, a schedule is not seen as a set of imposed obligations that need to be repeatedly adapted to the actual construction situation, but as a daily tool that helps planners monitor, organize and, if necessary, adapt the actual construction process.

Structuring the schedule

When formulating a schedule, planners should therefore structure it in a way that allows them to make effective and sensible adjustments and additions during the construction process. In large-scale projects, schedules are often confusing due to the range and complexity of the tasks involved. In such cases the individual tasks should be hierarchized using a clear structure of summary tasks. › Fig. 44 This enables the construction phases and workflows to be presented in detail and also to be seen within the overall deadline structure.

User-oriented schedule versions

This approach makes for easier comprehension of the schedule in its entirety. It can also be represented in different ways depending on the specific user. Typical modes of displaying the schedule include:

- Overview for project manager and clients: shows major summary tasks; does not show individual tasks
- Planning and contract award deadline schedule for the planning office: shows all individual planning tasks and award lead times: does not show construction tasks
- Construction schedule for site managers: does not show planning and award lead times; shows all contractor lead times and construction tasks
- Construction schedule as a guide for individual participants: shows the participant's tasks only

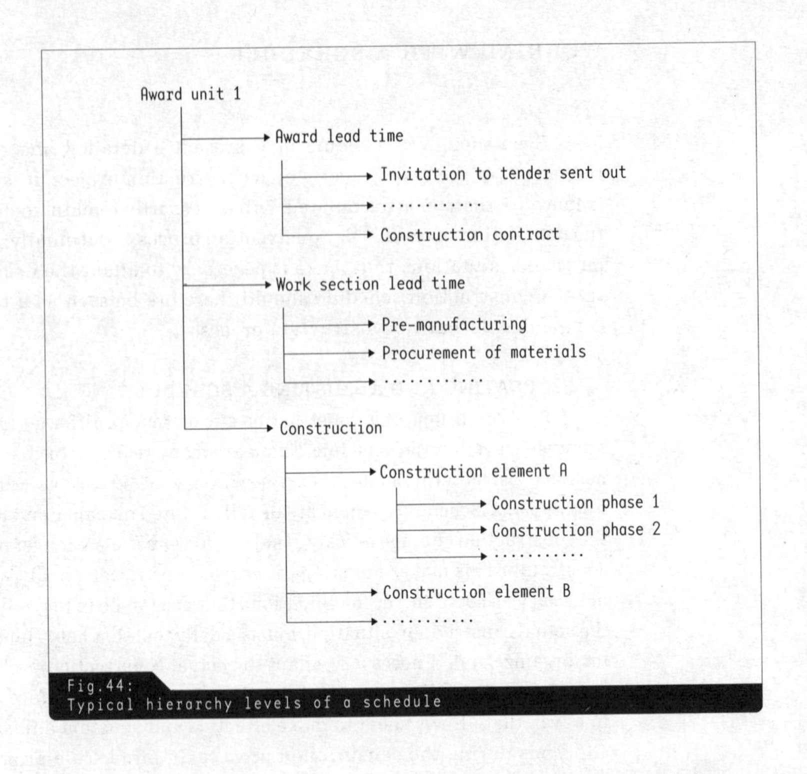

Fig.44:
Typical hierarchy levels of a schedule

By showing only those tasks that are relevant to a particular target group, the schedule provides a clear basis for participants in terms of planning and execution.

An important criterion for the relevance of all modes of display is that they be based on a single coherent schedule. If different schedules are used in parallel, the various users and multiple influences make synchronization difficult at a practical level. Integrating modifications from a single point and then passing them on to all participants using the hierarchical structure described above facilitates their application to the relevant parts of the overall process.

Integrating modifications

Apart from hierarchization, the usability of a schedule in practice is greatly improved when planners establish a clear connection between all sets of tasks. Identifying the effects of a modification requires the integration of all the tasks into a single context that allows the entire schedule to be updated automatically. At the same time it should be noted that not every delay or adjustment affects the deadline for completion.

Critical path

Usually there is only one dependency running through the project from start to finish—a dependency that, if subject to delays, has an immediate effect on the deadline for completion. This is referred to as the

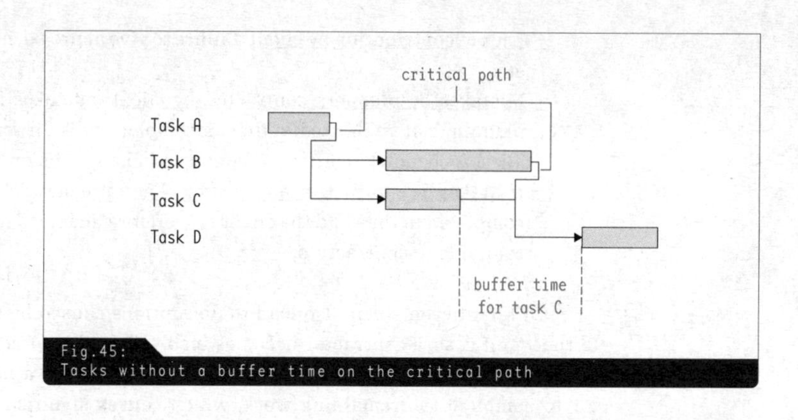

Fig.45:
Tasks without a buffer time on the critical path

critical path. In the case of other tasks, a buffer time can be used to prevent delays influencing the critical path. › Fig. 45

Buffer time

Every task that is not on the critical path has a buffer time that can be calculated by modern scheduling programs. This means that planners can directly monitor the degree of flexibility at their disposal when faced with a potential delay, and also see how much additional time they can allow the construction contractors involved. ›✎

DISRUPTIONS TO THE CONSTRUCTION PROCESS

Most of the modifications planners find it necessary to make to a construction schedule are based on disruptions to the construction process. Disruptions can be caused by clients (owners and planners and construction firms contracted by them), contractors or by third parties.

Disruptions from the client side

Typical examples of disruptions from the client side are:

_ Changes made by the owner: owner requests retroactive changes based on new user specifications, structural alterations to planning etc.

\\Tip:
Using colors, bar formats and hatchings to distinguish individual tasks, milestones, summary bars and entire construction areas improves a schedule's legibility. These features make individual work sections and construction phases easily identifiable. Automatic labeling of summary bars and tasks is also helpful.

\\Tip:
Calculating buffer time has an additional advantage. If there are no links between tasks in the subsequent phases of construction, the buffer time extends to the final project deadline. The scheduler can easily check longer buffer times to ensure that possible dependencies have not been overlooked.

_ Lack of contribution by client: failure to give approval, non-payment etc.
_ Mistakes by planners contracted by client: mistakes in planning, planning not submitted in time, incomplete calls for tenders, unrealistic schedule, insufficient construction supervision etc.
_ Mistakes by contractors engaged by client: preparatory work is not completed in time and the client is therefore unable to make the site available to contractors.

Disruptions by the contractor

Different events can also lead to disruptions caused by contractors. In the worst case, a contractor becomes insolvent and is forced to declare bankruptcy. The client is then forced to find and engage a new contractor to complete the remaining work, which causes significant delays to the construction process. On the other hand, construction firms that have taken on a large number of contracts often have problems meeting their contractual obligations with the workforce at their disposal. This can produce delays on the individual construction sites. Strikes and flu epidemics, for example, can also significantly reduce the number of staff that construction firms can allocate to a particular contract.

Capacity problems

Capacity planning is often also faced with problems. Construction firms plan their staffing capacities at regular intervals (e.g. weekly) and allocate their available workforce across different construction sites. Firms cannot usually vary the size of the workforce they allocate to individual construction sites from day to day. If schedules demand daily variations in workforce provision, there will be a high probability of disruptions. › Fig. 46

For this reason, in order to avoid problems at a later stage, planners should endeavor to schedule work sections in way that allows staff capacity to remain relatively constant.

Disruptions by third parties

Apart from clients and contractors, third parties contractually engaged in the construction process can also cause disruptions. These can

\\Tip:
In order to ensure that firms have a constant level of work, tasks are linked not only to other job areas but also to one another within the same work section. This allows schedulers to preplan a range of teams that can successively work on individual aspects of the same work section.

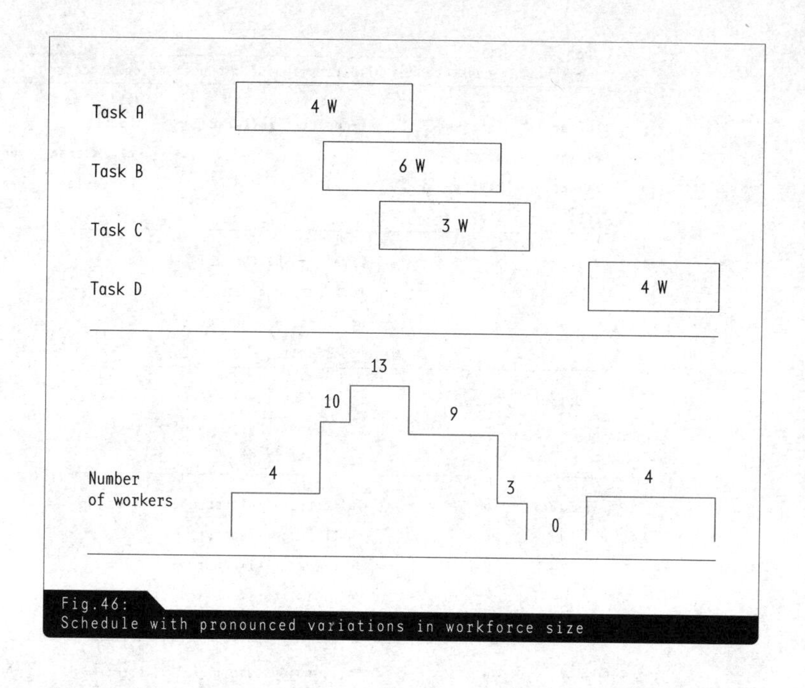

Fig.46:
Schedule with pronounced variations in workforce size

range from constraints and conditions imposed by authorities, strikes, and theft of clients' and contractors' property, to force majeure. If disruptions by third parties influence the client's situation, contractors have the right to claim an appropriate extension of the time scheduled for construction. If the influence of third parties affects the contractor's level of risk, the construction firm in question is nevertheless obliged to meet its obligations punctually. In the case of force majeure such as storm damage, flooding etc., the construction firm is usually granted a time extension. › Tab. 2

Effects of weather

Even in large-scale projects, unfavorable weather conditions in the winter months and other times of the year often lead to delays in meeting schedules. Although declines in performance over the winter months or in holiday periods can generally be offset by longer buffer times and task durations, predicting winter weather conditions precisely is impossible. Depending on the location of the construction site, frost and other adverse conditions can bring construction to a standstill for a long period. Early installation of heating or site-heating facilities can remedy this situation. However, planners need to note that, despite adequate temperatures inside the building, materials such as ready-mixed concrete, poured asphalt and ready-mixed plaster cannot be applied if external temperatures are too low.

Tab.2:
Contractual consequences of disruptions for construction process

Influence of the client (CL)	Influence of the contractor (CN)	Example of influence	Can CN claim time extension?	Can CL claim additional payment?
Direct	No	Directive by the CL, e.g. work stoppage due to lack of funds, or changes to construction	Yes	Yes
Indirect	No	Lack of CL involvement, e.g. late submission of approval	Yes	Yes
Indirect	No	Third-party influence on CL, e.g. preparation delays mean site not be ready for construction work	Yes	Yes
No	No	Force majeure, e.g. storm, war, environmental disaster	Yes	No
No	Indirect	In-house disruption to CN, e.g. flu epidemic or strike	No	No
No	Indirect	Third-party influence on CN, e.g. theft of equipment	No	No
No	Direct	Refusal to fulfill contract, e.g. too few workers on site	No	No

Types of disruptions

Disruptions relevant to the schedule can basically be divided into three categories: › Fig. 47

_ Delayed completion: A task begins at a later time but is completed within the prescribed task duration.
_ Extended construction time: A task requires longer than the planned task duration.

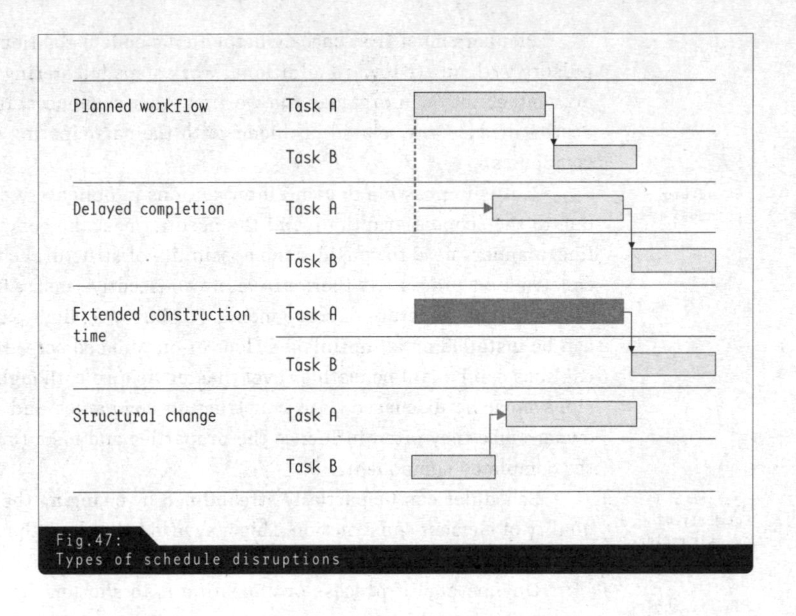

Fig.47:
Types of schedule disruptions

_ Structural change to the construction sequence: The construction process, or rather, the interdependent tasks, are arranged in a sequence that differs from the one planned.

DEALING WITH DISRUPTIONS

Should disruptions occur that require "critical path" intervention due to structural changes or because they could cause the project to miss its completion deadline, planners must endeavor to deal with these problems within the prescribed construction period. Possible forms of intervention include:

_ Checking necessary dependencies
_ Structural changes in the construction process
_ Changing the method or quality of construction
_ Shortening construction phases
_ Accelerating construction work

Necessary dependencies

Not every dependency is absolutely necessary, even if it makes sense. Work can sometimes be done retroactively if sufficient consideration is given to the subsequent tasks and if the connections to other components can be subsequently reworked.

Planners must first check whether a dependent relationship is compulsory with an eye toward additional work steps (plastering → painting), or whether they can organize the work in other sequences. It is probably best to discuss any related problems with the participating construction companies.

Structural changes

If it is impossible to avoid interventions in obligatory dependencies due to the scope disruptions and the need to meet the completion deadline, planners need to consider the possibility of structural changes to the construction process. If there are many consecutive tasks, it is possible to change the structure of components. For instance, light partition walls can be installed on an unfinished floor or on finished screed, and electrical lines can be laid beneath or over plaster. As a rule, though, these decisions must be discussed with construction companies and the building owner since they often influence the properties and visual impression of the completed component.

Method and quality of construction

Schedules can be partially streamlined by changing the method and quality of element construction. This may avoid long lead times in the premanufacturing process and long curing and drying times.

Shortening construction phases

One method of process optimization is to shorten the construction phases. As described above › Chapter Creating a schedule, The structure of the project schedule construction periods can be reduced if back-to-back work is organized into cycles. If the first contractor completes work on a floor only after the next contractor begins, more construction time is needed than if the floor is divided into smaller construction segments. In the latter case, one or several contractors can work on the floor at the same time. In case of doubt, a delayed contractor must be called upon to finish parts of the construction area in order to allow the subsequent contractor to start work there.

\\Example:
Work that produces a lot of dirt such as plastering, laying screed or putting in cut stone floors, should be scheduled before jobs such as carpeting and painting where surfaces are easily soiled. This is not to say that a stairway cannot also be laid with stone at a later stage in a project as long as areas with carpeting are closed off until the stone work has been completed and the stairway has been cleaned.

\\Example:
If planners are confronted with tight deadlines and wish to avoid curing and drying times for a cement screed and thus work holdups in the areas in question, they can, as an alternative, install poured asphalt or dry screed, which can be walked on one day later. It should however be noted that this is more expensive than cement screed.

Accelerating construction

In general, the client may ask a contractor to work more quickly, but distinctions must be made between the causes of the delay. If the contracting company is responsible, it must take all the necessary measures—including overtime and deployment of additional workers—in order to meet the agreed deadline, and it must do so on a cost-neutral basis. However, if the client or the site manager requests a third company that is not responsible for the delay to speed up work, the client must pay for any additional costs.

Preventing additional costs

Since the measures described above often produce additional costs, the owner must be involved in the decision-making process. It is ultimately the owner who must decide what funds he or she is prepared to mobilize to ensure that the building is completed on time.

Even in the initial phase of creating the schedule, planners are well advised to integrate delay periods. Problems almost always arise—insufficient preliminary work, delivery delays, theft of materials, etc.—and they will have to be dealt with. The time cushions in a schedule are an important way to ensure completion deadlines are met. If no cushion exists when planners create a schedule, it is usually a sign that the completion deadline is unrealistic.

Considering alternatives

Furthermore, at an early stage in the project, planners should also consider the latest point at which they will still be able to select an alternative construction method without violating the contract or incurring additional costs (e.g. cast-in-place concrete vs. prefabricated solutions, plaster vs. plasterboard, cement screed vs. dry screed, plaster flush with doorframes or closed frames). The deadline conditions should be analyzed in good time, and corresponding decisions discussed with the owner.

SCHEDULING AS PROCESS DOCUMENTATION

Scheduling is not only a method for organizing the construction process. It also performs an important function in documenting the project. Since it evolves over the entire planning and construction process, it can be used in retrospect to prove or disprove the occurrence of disruptions. This is important if there are unresolved claims between the owner and construction companies (e.g. compensation for damages) that need to be settled. Further, the schedules of completed projects are a source of data for future schedules and therefore represent important knowledge gained by the planning architect in the process.

Recording actual deadlines and disruptions

The main task of scheduling in this context is to record actual task durations as compared to the target task durations estimated by the scheduler. Disruptions and their causes should also be jotted down. One method is to document current events on the construction site by making handwritten entries into the current schedule. The paper copies, which should be filed at regular intervals, provide a basis for updating the schedule. Ideally, site managers will enter the deadlines directly into a scheduling program that will keep the schedule data constantly up to date. However, after each change, the previous version and its associated data should be archived under the proper date.

IN CONCLUSION

Complex building projects require a great deal of organization and coordination. Without solid scheduling, it is impossible to achieve effective time management of large construction projects. For both the architect in charge of this coordination and the site manager, it is extremely important to organize all the planning and construction processes in advance in order to remain in control of the situation. If these parties only respond to events and are unable to actively control the process, the self-organizational attempts of project participants will often result in disruptions, coordination problems, mutual interferences and delays.

Nevertheless, managing the planning and construction processes is not a matter of giving project participants written-in-stone deadlines that they must strictly follow. Planning specialists must consider all their concerns and integrate them into the management process in order to find solutions that everyone can implement.

A schedule is not merely a contractually agreed service between the building owner and the architect. It is also an effective instrument for the daily management of planning and construction processes. The creation of a realistic and implementable schedule involves effort, but keeping the entire planning and construction process in mind, planners will find that it makes later coordination and conflict resolution a good deal easier. It also lays the groundwork for short construction periods. The more architects devote themselves in advance to sequences in the construction process, the easier the work of site managers becomes.

LITERATURE

Bert Bielefeld, Thomas Feuerabend: *Baukosten- und Terminplanung*, Birkhäuser Verlag, Basel 2007

Tim Brandt, Sebastian Th. Franssen: *Basics Tendering*, Birkhäuser Verlag, Basel 2007

Chartered Institute of Building (ed.): *Planning and Programming in Construction*, Chartered Institute of Building, London 1991

Wilfried Helbig, Ullrich Bauch: *Baustellenorganisation*, Rudolf-Müller-Verlag, Cologne 2004

Hartmut Klein: *Basics Project Planning*, Birkhäuser Verlag, Basel 2008

Werner Langen, Karl-Heinz Schiffers: *Bauplanung und Bauausführung*, Werner Verlag, Neuwied 2005

Richard H. Neale, David E. Neale: *Construction Planning*, Telford, London 1989

Jay S. Newitt: *Construction Scheduling. Principles and Practices*, Pearson Prentice Hall, Upper Saddle River, NJ, 2009

Lars-Phillip Rusch: *Basics Site Management*, Birkhäuser Verlag, Basel 2008

Sandra Christensen Weber: *Scheduling Construction Projects. Principles and Practices*, Pearson Prentice Hall, Upper Saddle River, NJ, 2005

Falk Würfele, Bert Bielefeld, Mike Gralla: *Bauobjektüberwachung*, Vieweg Verlag, Wiesbaden 2007

Tab.3:
Information typically required by the main planner in the initial project phase

From:	**To:**	**Required information**
Structural engineer	Architect	_ Relevant structural systems and materials _ Full range of component dimensions
Building services engineer	Architect	_ Type of installations required for building use _ Location of utility and wiring rooms _ Route of main lines, required routes for main distribution lines _ Initial sizing of installations and lines
Architect	Structural engineer	_ Site plan, building form, floor height _ Maximum and most common width of columns _ Rough specifications
Architect	Building services engineer	_ Site plan, building form and size _ Use, user numbers (e.g. number of employees if used as office) _ Building services requirements _ Floor plans

Tab.4:
Information typically required by the main planner in the design phase

From:	**To:**	**Required information**
Structural engineer	Architect	_ Main and secondary axes of loadbearing elements _ Preliminary dimensions
Building services engineer	Architect	_ Initial sizing of installations and lines _ Openings necessary for building services _ Cost estimate
Architect	Structural engineer	_ Dimensioned design development drawings (plans and sections), ready to be submitted for the building permit
Building services engineer	Structural engineer	_ Location of the main lines, location and loads of the installations
Architect	Building services engineer	_ Final design development drawings (plans, sections, views)
Structural engineer	Building services engineer	_ Design of loadbearing structure (girders, columns, loadbearing walls) _ Openings and slits in loadbearing elements

Tab.5:
Information typically required by the main planner to prepare for construction

From:	To:	Required information
Structural engineer	Architect	_ Formwork drawings _ Reinforcement drawings _ Connection details _ Bills of material
Building services engineer	Architect	_ Electrical, ventilation, heating, sanitary planning _ Drawings of wall openings and slits for building services _ Tendering documents, e.g. main lines for the invitation to tender for the building shell _ Defined responsibilities for other planning specialists
Architect	Structural engineer	_ Updated dimensioned plans and sections _ Working drawings, design details, specifications _ Specifications from the preliminary building notification or the building permit
Building services engineer	Structural engineer	_ Location of the main lines, location and loads of the installations _ Drawings of wall openings and slits
Architect	Building services engineer	_ Approved plans and perhaps specifications from the authorities _ Specifications _ Construction drawings
Structural engineer	Building services engineer	_ Formwork drawings, steel structure drawings and timber structure drawings _ Location of steel reinforcement for wall openings

Tab.6:
Sample unit production times to roughly estimate task durations

Work	UPT	Unit
Preparing the construction site		
Setting up crane	10-50	h/unit
Steel-lattice fence	0.2-0.4	h/m
Connecting utilities (electricity, water)	0.2-0.5	h/m
Excavation		
Excavating building pit	0.01-0.05	h/m^3
Excavating individual foundations with power shovel, including removal	0.05-0.3	h/m^3
Excavating individual foundations by hand	1.0-2.0	h/m^3
Concrete		
Rough estimate for complete building shell (700-1400 m^3 gross volume and 3-5 workers)	0.8-1.2	h/m^3 GV
Binding layer, unreinforced, d=5 cm	0.2	h/m^2
Bottom slab, reinforced cast-in-place concrete, d=20 cm	2.0	h/m^2
Ceiling, reinforced cast-in-place concrete, d=20 cm	1.6	h/m^2
Precast and partially precast concrete ceilings	0.4-0.9	h/m^2
Entire building, prefabricated	0.3-0.7	h/t
Casting concrete elements (without formwork or reinforcement)	0.4-0.5	h/m^3
Casting walls (without formwork and reinforcement)	1.0-1.5	h/m^3
Casting columns (without formwork and reinforcement)	1.5-2.0	h/m^3
Cast-in-place concrete stairway (without formwork and reinforcement)	3.0	h/unit
Large-panel formwork	0.6-1.0	h/m^2
Single formwork	1.0-2.0	h/m^2
Reinforcement	12-24	h/t

All types of sealing	0.25-0.40	h/m^2
Scaffolding (assembly and disassembly)	0.1-0.3	h/m^2
Masonry		
Loadbearing masonry wall	1.2-1.6	h/m^3
Non-loadbearing interior wall	0.8-1.2	h/m^3
Carpentry work		
Rafter roof, including joining and mounting (based on roof area)	0.5-0.7	h/m^2
Roofing		
Flat roof (gravel), including complete mounting of non-insulated roof	0.5-0.7	h/m^2
Pitched roof with roofing tiles	1.0-1.2	h/m^2
Metal roofing	1.3-1.5	h/m^2
Cladding for exterior walls		
Metal facade cladding	1.0-1.3	h/m^2
Facing brick leaves	1.1-1.5	h/m^3
Composite heat insulation system	0.6-0.8	h/m^2
Assembly of precast concrete facades	0.5-0.7	h/m^2
Exterior wall cladding with natural stone, slate, etc.	0.5-0.8	h/m^2
Window construction		
Installing individual windows	1.5-2.5	h/unit
Installing roller shutter housing	0.6-1.5	h/unit
Roof windows	2.5-3.5	h/unit
Interior window sills	0.3-0.5	h/m
Plaster		
Exterior plastering	0.5-0.7	h/m^2

Interior plastering, done by machine	0.2-0.4	h/m²
Interior plastering, manual	0.3-0.6	h/m²
Ceiling plaster	0.3-0.4	h/m²

Screed

Laying cement screed and anhydride screed (without membranes, insulation, etc.)	0.1-0.3	h/m²
Laying mastic asphalt screed (without membranes, insulation, etc.)	0.3-0.5	h/m²
Floating floor screed, including insulation layer	0.6-1.0	h/m²
Terrazzo screed, polished	2.0-2.5	h/m²

Dry construction

Drywall with plasterboard	0.2-0.5	h/m²
Prefabricated walls or wall paneling, single layer, including substructure	0.7-0.8	h/m²
Covering slanted ceilings	0.3-0.5	h/m²
Suspended ceiling structures	0.6-1.1	h/m²
Plasterboard stud wall, single panel	0.4-0.8	h/m²
Plasterboard stud wall, double panel	0.6-1.5	h/m²

Doors

Installing steel frames + door leaves	1.9-2.5	h/unit
Installing wooden doors	1.0-1.5	h/unit
Exterior doors	2.5-4.5	h/unit

Tiles, paving stones, cut stones

Floor tiling	0.5-1.8	h/m²
Wall tiling	1.3-2.5	h/m²
Natural and concrete paving stones	0.8-1.2	h/m
Baseboard made of tile or natural stone	0.3-0.4	h/m

Flooring		
Creating level surface, filling holes	0.05-0.2	h/m^2
PVC, linoleum and rolled flooring materials	0.3-0.6	h/m^2
Needle felt or carpet on screed	0.1-0.4	h/m^2
Baseboards	0.1-0.2	h/m^2
Parquet floors, including surface treatment	1.2-1.8	h/m^2
Sanding parquet floors, surface treatment	0.2-0.3	h/m^2
Natural stone floors	0.9-1.2	h/m^2
Stairway coverings	0.5-0.7	h/m^2
Painting and wallpapering		
Putty work	0.1-0.2	h/m^2
Standard wallpaper (wall chip wallpaper, thick embossed wallpaper, etc.)	0.1-0.4	h/m^2
Special wallpaper (velour, textile, wall images, etc.)	0.3-0.8	h/m^2
Painting interior wall, single coat	0.05-0.2	h/m^2
Painting interior wall, three coats	0.2-0.5	h/m^2
Plastering and painting exterior wall	0.2-0.8	h/m^2
Painting window, per coat	0.2-0.6	h/m^2
Painting metal surface, all required coats (doors, sheet metal walls, etc.)	0.3-0.6	h/m^2
Painting metal elements, all required coats (frames, sheet metal covering, etc.)	0.6-1.0	h/m^2
Painting metal railings	0.1-0.3	h/m
Electrical work		
Rough estimate for all electrical installations (700-1400 m^3 gross volume and 2-3 workers)	0.2-0.4	h/m^3 GV
Assembling cable tray + electrical lines	0.3-0.5	h/m
Assembling lights	0.3-0.8	h/unit

Assembling sub-distribution board	0.5-1.0	h/unit
Detailed installation of switches, outlets, etc.	0.02-0.05	h/unit
Heating, plumbing and sanitation installations		
Rough estimate for complete heating installation (700-1400 m^3 gross volume and 2-3 workers)	0.1-0.3	h/m^3 GV
Rough estimate for complete gas, water and wastewater systems (700-1400 m^3 gross volume and 2-3 workers)	0.15-0.4	h/m^3 GV
Rough assembly of pipe routes	0.4-0.8	h/m
Rain and wastewater pipes	0.10-0.50	h/m
Detailed installation and assembly of sanitary fixtures	0.3-1.0	h/unit

THE AUTHOR

Bert Bielefeld holds a doctorate in engineering and works as a freelance architect in Dortmund. He is the managing director of the Aedis Pro-Manage project management company, teaches construction economics and construction management at the University of Siegen, and lectures at various architectural chambers and associations.

导言

将一个设想转化成一幢实际的建筑是一个非常复杂和漫长的过程。在这个过程中将涉及许多不同身份的人——建筑承包商、建筑设计师以及建设方（业主）等——所以有必要在这个过程中将不同人的工作进行协调统一。

建筑师和工程设计人员为业主代理技术方面的事务，他们必须确保整个工程期间的各项工作顺利地进行。为了确保业主的利益，在设计阶段建筑师和工程设计人员需要协调不同身份参与者之间的工作，在施工阶段则需要在施工现场对建筑承包商的工作进行监督。对于较大的工程，在设计和施工阶段常常需要 20～30 名建筑师以及工程设计人员参与其中，这就导致不同人员之间存在着复杂的联系和相互依赖关系。对于每一个设计人员而言，一般很难明确或者判断自己所担任的工作与整个工作流程之间的联系。由于建筑师的设计工作包含了整个工程的各项工作，故他们具有更加特殊的责任，需要对不同的工作者之间进行协调。也就是说，他们是整个工程中唯一把握“整张蓝图”的工作者。

编制施工进度计划是用来控制整个工程活动过程的工具。本书对施工进度计划的基本原理和应用进行了阐述，并对其中具有代表性的工序的表达形式和表达深度进行了着重讲解。本书的目标是向学生们对施工进度计划进行一个简洁的结合实际工程的讲述。然而，在施工进度计划编制完成之后，所需进行的协调工作并没有结束。还需要我们对这个工程进行不断地更新和修改，以确保我们的进度计划更加精确。在进行现场施工之前，需要我们做大量的准备工作，并对可能的细节进行深入考虑，以确保编制的施工进度计划能够准确地指导现场的工作。本书接下来的章节内容将讲述编制进度计划工程需要考虑哪些工作者以及哪些施工过程。

进度计划的创建

进度计划的基本要素

首先，让我们介绍一些主要的术语和进度计划中的不同基本要素。

工期和截止期

工程设计人员需要分清工期和截止期的不同。“截止期”指的是一个特定的时间点，比如某一天工程中某项工作必须完成；而“工期”指的是一段时间（比如在14天内完成某一项工作）。

任务

任务是进度计划中最基本的要素，指的是独立的工作单元（比如，为地面铺设瓷砖）。如果是多项任务组合在一起（比如抹灰并铺设瓷砖），便形成了汇总性任务。〉参考“进度计划的创建”一章中“工程进度计划的结构”部分内容

确定任务持续时间和施工顺序

任务持续时间指的是完成一项任务所需要的时间，它是决定工作量和生产效率的因素之一。〉参考“进度计划的创建”一章中“工程进度计划的结构”部分内容

任务持续时间确定主要是对不同任务的持续时间进行计算分析，而确定不同任务之间的联系和依赖关系则是任务顺序规划。将二者组合在一起，则构成了施工进度计划的基础。〉参考图1

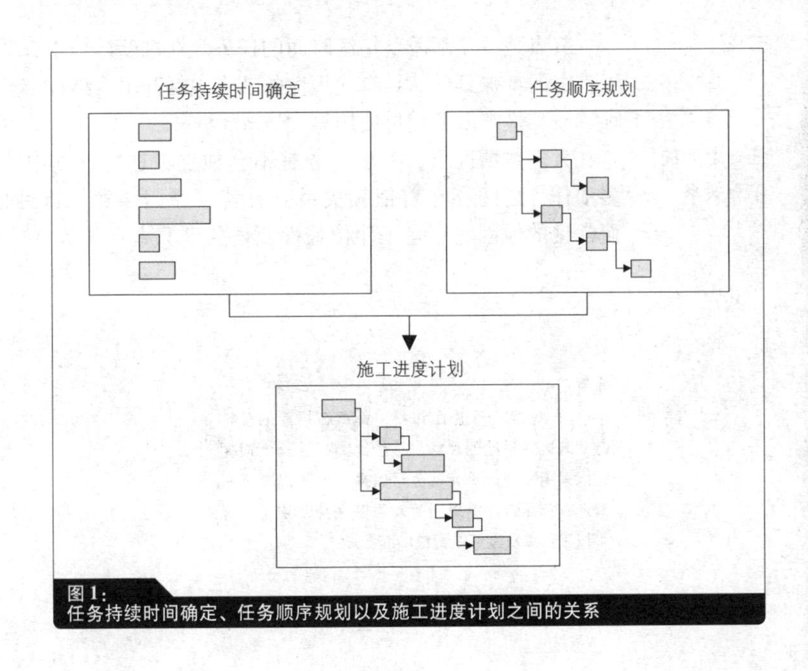

图1：
任务持续时间确定、任务顺序规划以及施工进度计划之间的关系

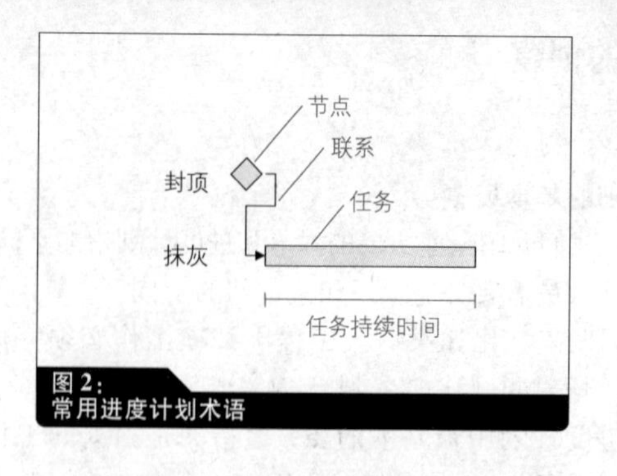

图 2：
常用进度计划术语

施工方法和资源

施工方法指的是完成一项任务所采用的技术过程。

资源指的是进行一项任务所需要的设备器材和劳动力等。在一项建筑工程的准备过程中，建筑公司需要对其资源进行计划安排并准确计算施工成本，以便于确定相应的施工方法。该分析结果是建筑公司用来投标的根本依据。然而，建筑师在编制进度计划时并不会着重考虑施工方对资源计算分析。为了确保连续的工作流程，则需要对任务持续时间进行一个符合实际的估算，而资源分析能够为估算结果提供一个依据。〉参考“进度计划的创建”一章中“表达深度”部分内容

节点

节点是一项不需要持续时间的任务。在进度计划中，节点是独立于其中的特殊事件。具有代表性的节点包括开工、结构部分完工、封顶、竣工检查以及交付使用等。〉参考图 2

任务之间的依赖关系

大多数情况下，一项任务是不会独立在进度计划中的，而是与多项任务之间存在着依赖关系。有若干原因导致了这种依赖关系，而常见的原因之一是工序依赖性：任务 B 只有在任务 A 完成之后才

举例：

对于任何一个施工过程，都存在可能的多种方式来达到预期的目标。打个比方，混凝土顶棚可以采用预制板单元，也可以采用现浇混凝土来建造；墙面砖可以采用薄灰浆贴在抹灰层上，也可以采用厚灰浆贴在粗糙墙面上。

可以开始（打个比方：地面层墙体施工→地面层顶板施工→二层墙体施工）。

另外，有一些任务只有在和其他任务平行进行时才可以完成（比如，在为多层建筑搭设脚手架的时候，脚手架只能随着建筑的升高而升高）。但是，这种过程依赖性往往可以通过进一步的细节分析转化为顺序依赖性。

与之相反的是，对于装修承包商而言，一些施工过程往往无法平行进行（比如，砂浆铺设和抹灰）。我们称这种情况为一项任务干扰另一项任务。这也正是设计者进行施工过程依赖性分析以及必要时将工程分为不同施工阶段进行施工的重要原因。〉参考“进度计划的创建”一章中“工程进度计划的结构”部分和“设计和施工过程中的工作流程”一章内容

关系类型

不同的关系类型在表达两项任务之间的依赖性时具有非常重要的作用。在施工进度计划中存在四种不同的关系类型〉参考图3：

— 完成－开始：任务B只有在任务A完成之后才能开始。这种关系类型是最常见的形式，可以应用在诸如内墙砌筑（A）和内墙抹灰（B）两项任务之间；
— 完成－完成：任务A和任务B必须同时完成。当任务A和任务B需要为另外的任务提供条件的时候存在这样的关系。比如：由于室内任务需要确保密闭的空间，那么需要安装窗户（A）和封闭屋顶（B）；
— 开始－完成：任务B必须在任务A开始的时候完成。在这种关系类型中，为了防止某项任务干扰另一项任务，可以将其安排在可能的最晚完成时间点之前；

提示：

如果在进行施工进度计划编制的时候没有认真考虑各项任务之间的依赖性，那么往往会导致施工过程的中断和延长。比如，在进行残疾人钢框架门安装的时候，需要在地面抹灰之前完成插座的安装。如果忽视了类似任务之间的依赖性，可能会导致需要在已经完工的地面上再进行额外的工作。

注释：

常用的施工进度计划编制程序都支持上述的几种关系类型。通常情况下，它们对每项任务进行编号，然后对号码添加相应的依赖关系。比如，如果某项任务需要在5号任务之后完成，那么这样任务就被标记成5FTS，FTS指的是两项任务之间所存在的“完成－开始”的关系类型。

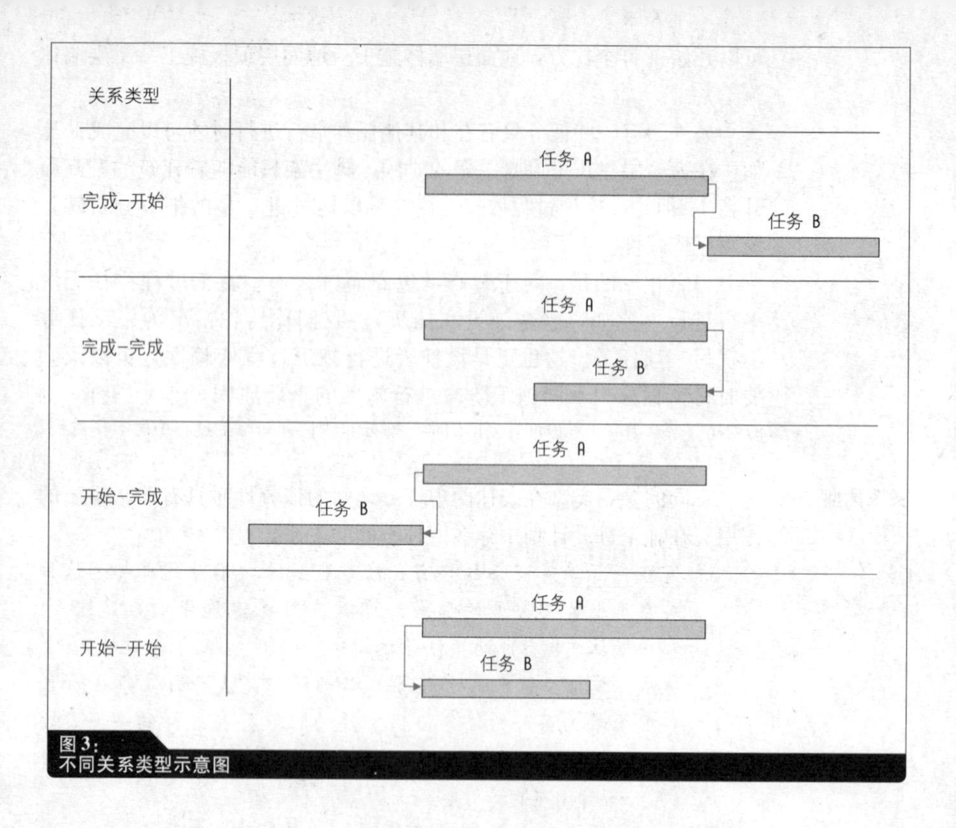

图 3：
不同关系类型示意图

— 开始 - 开始：任务 A 和任务 B 必须同时开始。当两项任务平行进行的时候，可能存在这样的关系类型，比方说，某一承包公司的工人想借用另一家承包公司的吊车进行大型构件搬运。

P16

表达形式

进度计划存在多种不同的图表表达形式。为了明确而有效地表示进度计划中的内容，可根据图表的目的和用途将其分为以下几种不同表达形式 〉参考图 4：

— 柱状图；
— 线形图；
— 网络图；
— 期限表。

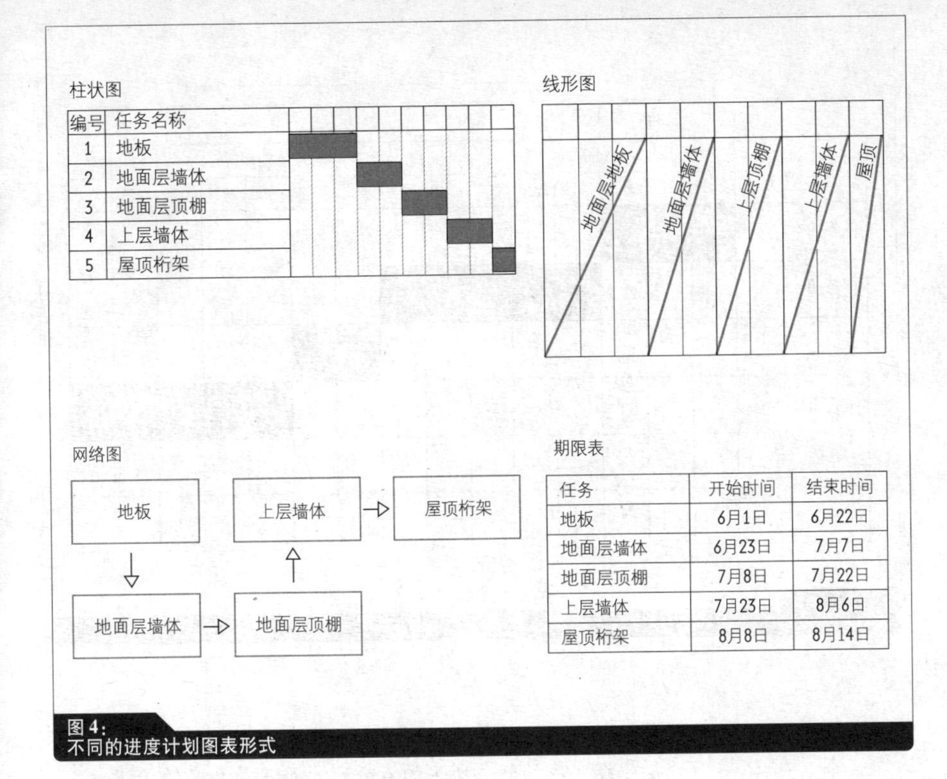

任务	开始时间	结束时间
地板	6月1日	6月22日
地面层墙体	6月23日	7月7日
地面层顶棚	7月8日	7月22日
上层墙体	7月23日	8月6日
屋顶桁架	8月8日	8月14日

图4：
不同的进度计划图表形式

柱状图

在建筑施工中，进度计划一般采用柱状图（也称为甘特图）进行表示。其中，水平轴为时间轴，而不同的任务则沿竖向轴分布。不同任务的持续时间则用与时间轴平行的水平柱状图进行表达。〉参考图5

根据工程规模和所需细节程度的不同，时间轴的单位可以是月、周或者天。图表中除了对任务进行表达之外，还可以在任务栏的左边增加任务的相关信息以增强图表的可读性（比如，任务描述、开始日期、结束日期、任务持续时间以及在必要的时候说明与其他任务之间的依赖关系）。柱状图中，任务之间的依赖关系一般采用箭头进行表示。

线形图

线形图与柱状图在基本结构上存在不同。线形图中的坐标单元列在时间轴下面的第二轴线上，任务则采用坐标系统中的直线描出。在建筑业中，常用的线形图主要有以下几种形式：

— 空间－时间图：采用几何线段表示完成的工程量（比如，表示公路施工中的一段）；

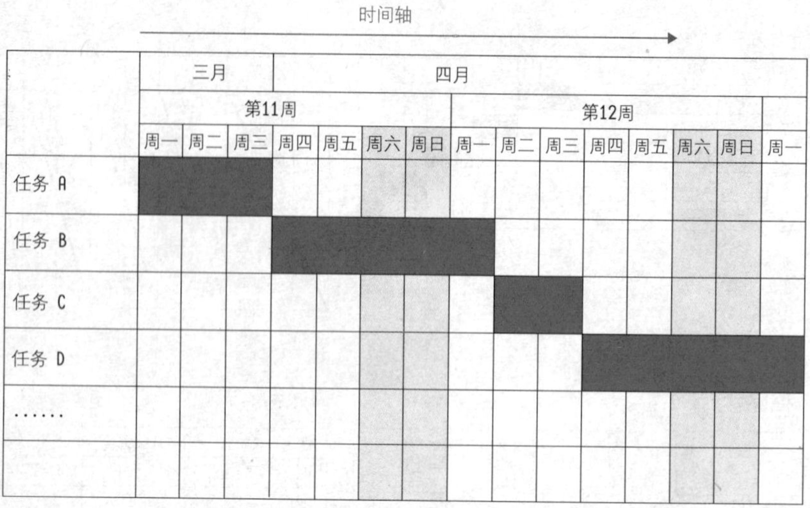

时间轴

	三月			四月											
	第11周							第12周							
	周一	周二	周三	周四	周五	周六	周日	周一	周二	周三	周四	周五	周六	周日	周一
任务 A															
任务 B															
任务 C															
任务 D															
……															

图 5:
柱状图的原理

— 数量 - 时间图：将所有工程量归一化成为 100%，在图中表示的是已经完成的任务所占的百分比，而不是真实的工程量。

线形图在建筑工程中使用得并不多，而是更多地使用在具有线性施工场地的工程中，比如街道、隧道以及下水道系统中。在这些工程中，不同的施工过程之间存在规则的循环关系。而在建筑施工中，许多任务（特别是装修承包商所进行的工作）都必须采用平行的方式进行表达，在线形图中将不能清楚地进行表示。然而，线形图能够更清楚地对目标性能进行比较。〉参考图 6

网络图

网络图采用网络的形式取代沿着坐标轴列表的形式来表示进度计划。通过网络图，设计者可以有效地制定出任务之间的相互关系，但是却只能表达出整个过程中有限部分的顺序关系。

网络图可以分为以下三种不同形式，其中最常见的是节点表示活动的形式：

— 节点表示任务：采用节点表示任务，采用箭头表示依赖关系；
— 箭头表示任务：采用箭头表示任务，采用节点间的连线表示依赖关系；

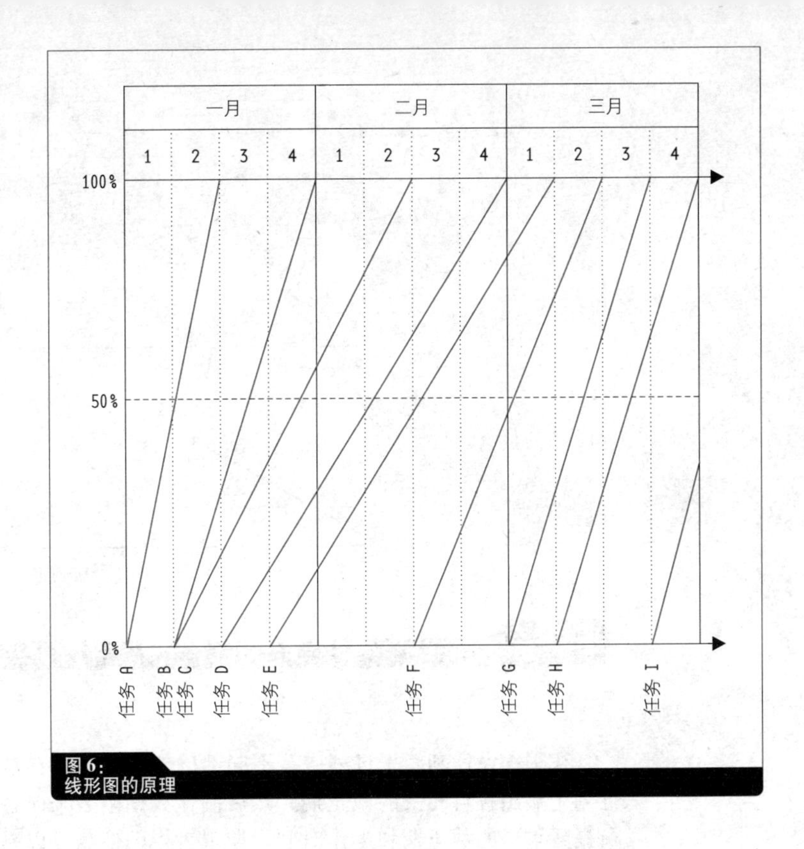

图 6：
线形图的原理

— 节点表示事件：采用箭头表示依赖关系，采用节点表示结果（不包括持续时间）。〉参考图 7

在施工进度计划中一般很少使用网络图进行表示，但在高级施工进度计划编制软件中一般都具有网络图功能，用来作为柱状图的补充。因此，网络图为进度计划编制增加了一种重要功能。相较于柱状图，网络图表示任务之间依赖关系的功能更加强大。通过在两种图表之间相互转化，编制者能够更好地了解和熟悉进度计划。任务的记录和持续时间可以通过柱状图来表达，而任务之间的依赖关系则可以表示在网络图中。

期限表

期限表是一种非常简单的表示形式。在施工进度计划中，一般使用期限表来表达一些重要的期限和时期。所以，期限表仅仅能够表示整个工程中的有限信息。期限表中所列举的截止期越多，其可读性越差，工程人员在理解各个截止期之间的关系时越困难。

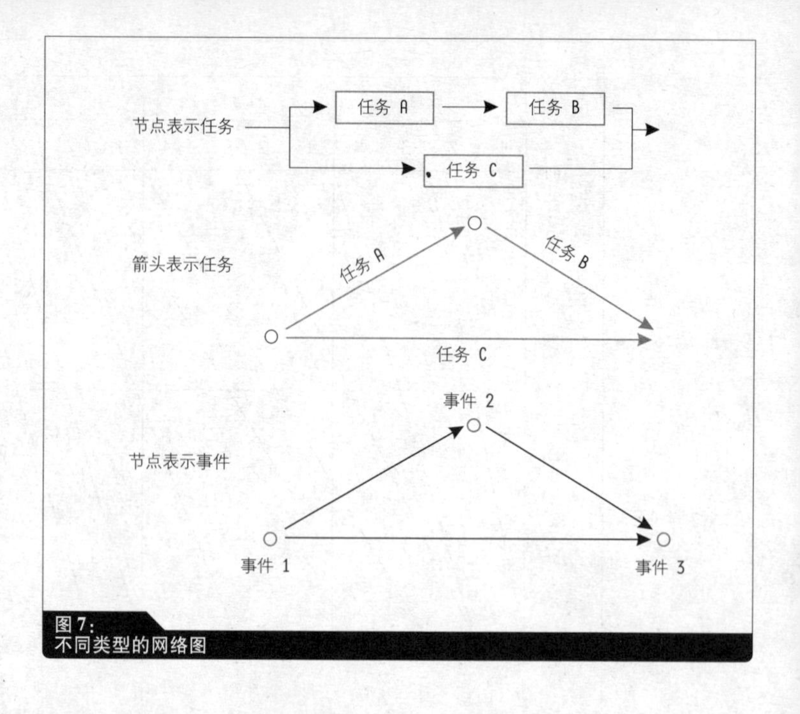

图 7:
不同类型的网络图

在工程设计和施工过程中，不同部门的工程人员会从整个进度计划中摘出各自相关的截止期，然后制作成期限表来向各自团队成员提醒重要的截止期和工作时间。期限表中的信息可以是相互联系的，打个比方，设计人员所需要的设计时间与合同规定中所规定的每个承包商的工期等。在投标的时候期限表一般随标书同时提交，中标之后该期限表可能会写入施工合同中作为合同条款。很多进度计划编制软件都可以从整个进度计划中提取出任务截止期，并存入单独的文件中。

P20

表达深度

进度计划应该具有一定的表达深度，以达到工程所需要的明确性、实用性以及详细程度。根据看待工程的角度不同，表达深度的要求会存在非常大的不同。一般来说，对于施工过程的表达存在以下三个不同深度：

— 从业主的角度：通过制定框架计划确定截止期；

— 从设计方的角度：利用工程进度计划对工程参与者进行协调；

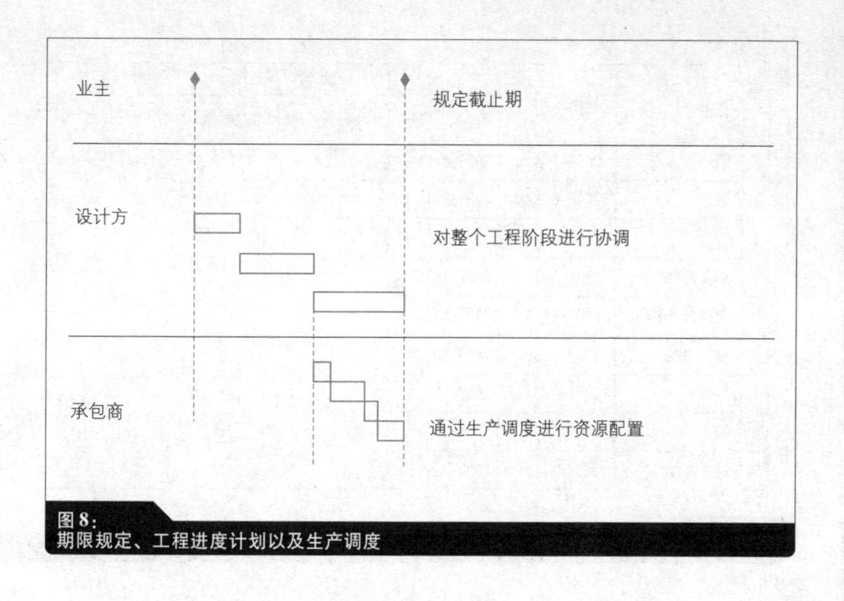

图 **8**：
期限规定、工程进度计划以及生产调度

— 从承包商的角度：通过生产调度进行劳动力和资源配置。〉参考图 8

另外，进度计划还可以根据时段（短期、中期或者长期）、编制者（业主或者承包商）以及细节深度（初步的、详细的或者非常详细的）进行分类。

框架计划

业主一般对于建筑物什么时候能够或者必须交付使用有一个明确的意见。某个百货公司可能需要在圣诞节之前完工，也可能业主已经和租户签订了租约，需要在某个具体的日期之前交付使用。设计者必须考虑业主所规定的截止期。一般情况下，工程的财务资助方（银行）也可以参与截止期的规定。为了符合业主对于截止期的规定以及大致把握整个工程的框架，设计者编制了框架计划来把握整个工程的概况。框架计划中覆盖了广泛范围的任务以及设计过程和施工过程大致的截止期。常见的任务包括：〉参考图 9

— 项目准备；
— 设计；
— 建设工程规划许可证申请；
— 施工准备；
— 开工；
— 结构施工；

编号	任务	开始日期	完成日期
1	项目开发	2008-7-4	2008-8-11
2	设计	2008-8-17	2009-1-1
3	建设工程规划许可证	2008-12-12	
4	施工准备	2009-1-1	2009-1-15
5	建筑外壳施工	2009-1-15	2009-5-31
6	建筑装修	2009-4-30	2009-10-30
7	设施安装	2009-7-31	2009-11-11
8	竣工	2009-11-11	
9	竣工检查	2009-11-15	2009-12-15

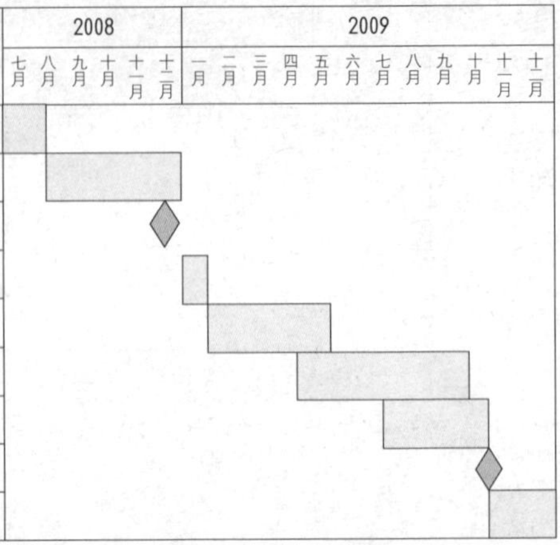

2008：七月 八月 九月 十月 十一月 十二月

2009：一月 二月 三月 四月 五月 六月 七月 八月 九月 十月 十一月 十二月

图 9：
框架计划图例

— 外壳施工；

— 各种装修；

— 竣工。

工程进度计划

工程进度计划一般是由建筑师编制的，其主要作用是用来协调建筑设计阶段和施工阶段中的所有参与人员。为了能够有效地组织和协调各种不同的设计活动，工程进度计划必须比框架计划更加详细。〉参考图 10 根据不同任务之间的依赖性，可以对任务进行组合和拆分。打个比方，对于建筑外壳施工，由于其通常是由同一个承包商来施工的，所以并不需要制定非常详细的施工计划。相反，干式墙的施工过程可能包括电气安装、卫生设备安装、门窗安装以及涂装工作等，所以有必要将其划分成若干施工过程。〉参考“设计和施工过程中的工作流程”一章中“装修工作”部分内容如前文所述，最有效的工程进度计划的详细程度是和工程的复杂程度、时间框架相对应的。

此外，工程进度计划应该全面地包括工程的设计和施工阶段。理想情况下，工程进度计划的编制应该能够让不同的工程参与人员（包括施工图设计人员、投标人员以及施工管理人员）都能够非常清晰和准确地从其中查阅与各自相关工作的截止期。〉参考“设计和施工过程中的工作流程”一章中“设计协调”部分内容

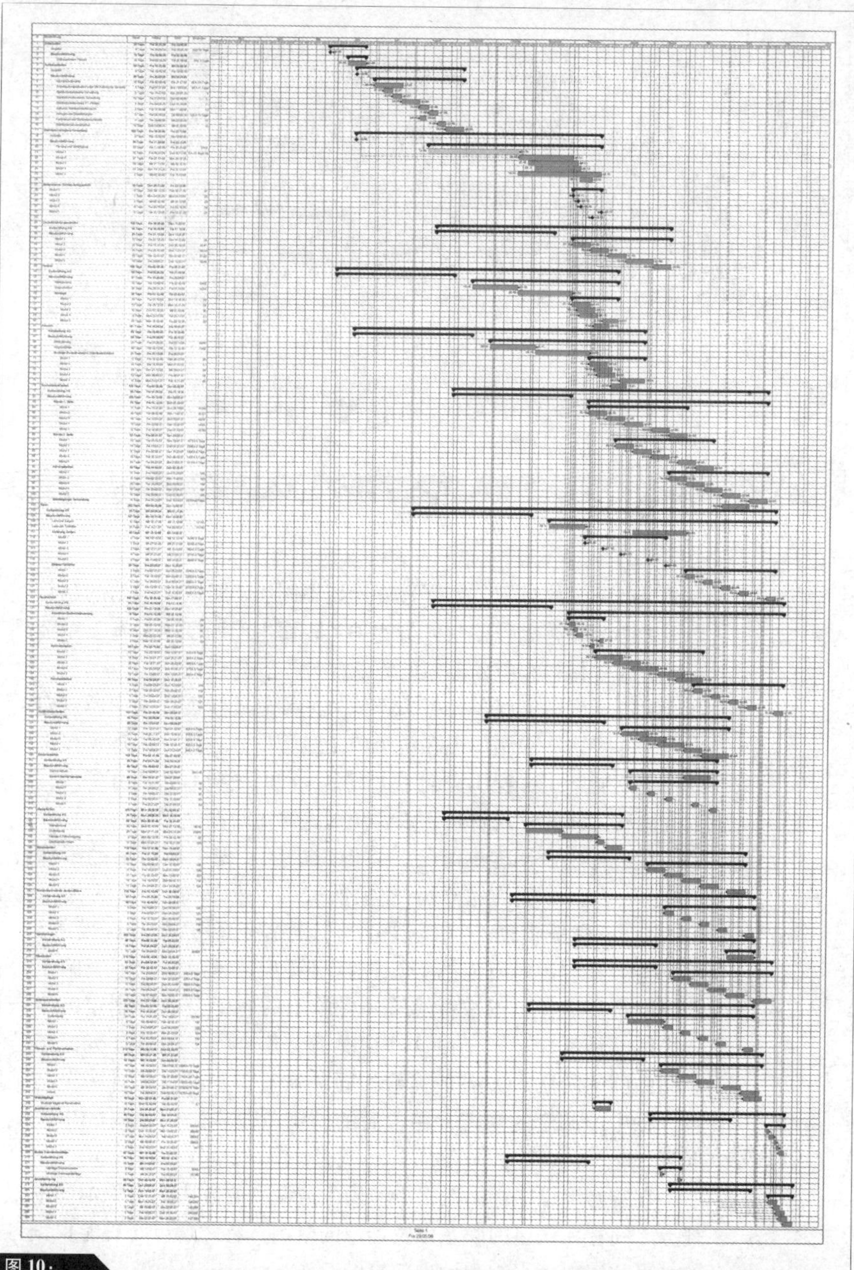

图 10：
工程进度计划示例

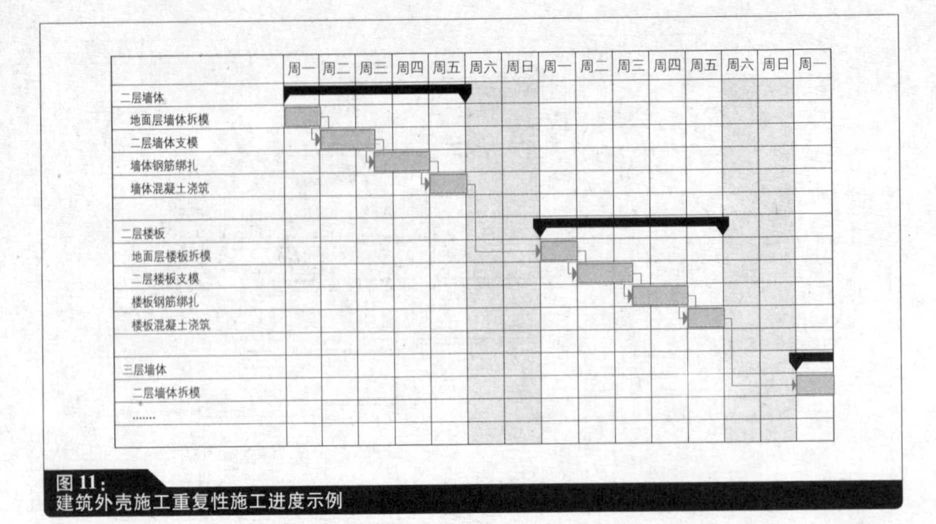

图 11：
建筑外壳施工重复性施工进度示例

生产调度

生产调度的目标与工程进度计划不同。工程进度计划的制订主要是用来协调所有工程参与人员的关系，而生产调度则是施工企业用来安排工人、材料和施工器械的方法。

生产调度的实施需要考虑框架计划和施工进度计划中所规定的截止期，并将这些限制体现在每个施工过程中。在进行生产调度时，需要确定不同施工过程所需的工人数量，同时还要有效地分配所需的施工机械和足够的建筑材料，以避免对施工过程产生影响。

在进行生产调度的时候，将施工进度计划中所给出的工作任务（打个比方，地面层混凝土楼板的安装）拆分成单独的施工过程（支模、绑楼板钢筋、浇筑混凝土、混凝土养护以及拆模等）。同时，生产调度还对所需的资源进行了调配。〉参考图 11 由于施工过程的周期重复性，负责进行建筑外壳施工的施工企业所制定的生产计划往往是重复性计划。在这种情况下，整个施工项目被分成了若干相同的施工阶段。如果工程量相同，施工企业只需要对一个施工阶段进行生产调度的制定，在随后的施工阶段中采用相同的生产调度即可。

在装修工程中，一般很少使用生产调度的方法。由于不同的施工任务之间存在太多的联系以及场地的限制，装修承包商只能根据各自的水平和实力来组织和计划各自的工作。

P25

框架计划的编制

通常情况下，框架计划是在综合考虑业主所期望的截止期和所有工程参与人员的可能完场时间的基础上编制的。对于大型项目，如果业主有自己的项目经理或者相关专业人员，那么通常由业主制定框架计划，然后将该计划作为限制因素交给设计人员。

业主对截止期的限定

业主对截止期的规定拥有无可替代的优先权，一般将直接规定出完工日期或者通过要求施工期限的间接方式（打个比方，要求在本财政年度内开工以减少税金的缴纳等）进行规定。从设计开始到建筑完工之间的时间段给出了整个设计和施工阶段的时间框架。

将整个时间段划分为设计阶段和施工阶段

将整个时间段划分为设计和施工两个阶段是非常关键的一项工作，它将决定工程人员是否能够在规定的时间内将设计和施工两个阶段均顺利完成。在施工阶段，可以通过提高工作效率等优化方式对工期进行缓冲，同时也有可能对设计阶段的工作在时限上进行优化，但这些优化工作都必须满足一定的限制条件。设计和施工两个阶段都需要满足一些基本要求，而且所给的工作时限一般并没有太大的富裕，所以很容易有超过期限的现象发生。

项目在动工之前需要进行一系列的工作：取得开工许可证、进行可行性评估并获得认可、与施工单位签订施工协议等。其中与施工单位的施工协议是以设计施工图、招标文件以及施工单位的投标文件作为先决条件来签订的。

一般来说，施工工作无法较大程度地偏离已有的标准施工过程；同样在进行施工流程组织设计的时候也存在类似的限制。

如果业主规定了截止期，那么相关的工期需要根据该截止期进行反算，或者根据类似的工程进行估算。通过这种方式，工程设计人员需要确定工程开工的日期，并确定工程开始时间和开工时间之间的时间段是否能够满足设计过程的需要。即使是在参考其他类似工程的时候，工程设计人员也必须时刻将工程的复杂性考虑在其中。

如果有依据表明某工程方案确实不可行，那么就必须考虑相应的替代方案。打个比方，采用替代的施工方法可能会缩短施工时间（采用预制构件或者采用所需干燥时间较短的建筑材料等）。如果采取了这些措施之后还无法在预期的截止期内完成工程，那么应该尽早和业主就这些问题进行沟通。

任务组织

框架计划除了将整个工程划分为设计阶段和施工阶段之外，还包含了一些关键的设计步骤和施工阶段，并将它们作为单独的任务或者节点表示出来。〉参考“进度计划的创建”一章中“进度计划的基本要素”部分内容框架计划对整个工程进行了一个概述，并且确保所有的工程人员能够按时完成各自的工作。工程进度计划则较框架计划更加详细，但是二者的界限并不是固定的，而是根据业主对信息的要求能够随时改变的。

P26

工程进度计划的结构

由于工程进度计划的作用是协调工程参与人员，所以其结构也是根据不同工程人员的工作内容而确定的。工程进度计划用工程任务将不同的工程设计人员和施工企业联系成为一个整体。

根据合同工作包制定进度计划层次纲要

为了建立工程进度计划的纲要和层次，需要将单独的任务组合成为汇总任务。打个比方，单独的任务可以划分到构件组施工或者某个施工阶段中。〉参考图 12

进度计划的最高层次应该是响应的合同工作包。“工作包”指所签订的设计合同或者施工合同中所规定的需要完成的建筑任务。如果某一工作包中涉及了多个承包商，其中的每一个承包商均要遵守合同工作包的内容。这种做法的优点在于承包商在工程的早期阶段就可以了解各自的工作内容和截止期。另外，不需要参考其他资料，仅仅根据合同内容就可以明确不同工作的截止期并制定相应的进度计划。〉参考“设计和施工过程中的工作流程”一章中“施工准备工作的协调”以及“施工进度计划表的使用”一章中“施工进度计划表的更新与调整”部分内容

举例：

与一个需安装精密仪器的小规模图书馆工程相比，对于一个简单的门厅工程，我们能够更加容易地估计到相应的设计内容和施工工期。因为对于该图书馆工程而言，需要更多工程设计人员的参与，相应需要进行更多的协调，也使工程受到干扰的可能性增加。

注释：

在大多数进度计划编制软件中，设计者可以通过在低层次任务中添加汇总任务的方法来形成进度计划的纲要。此时，所对应的低层次任务自动成为汇总任务，其任务持续时间则可以通过相应下属任务的工作时间进行叠加计算得到。

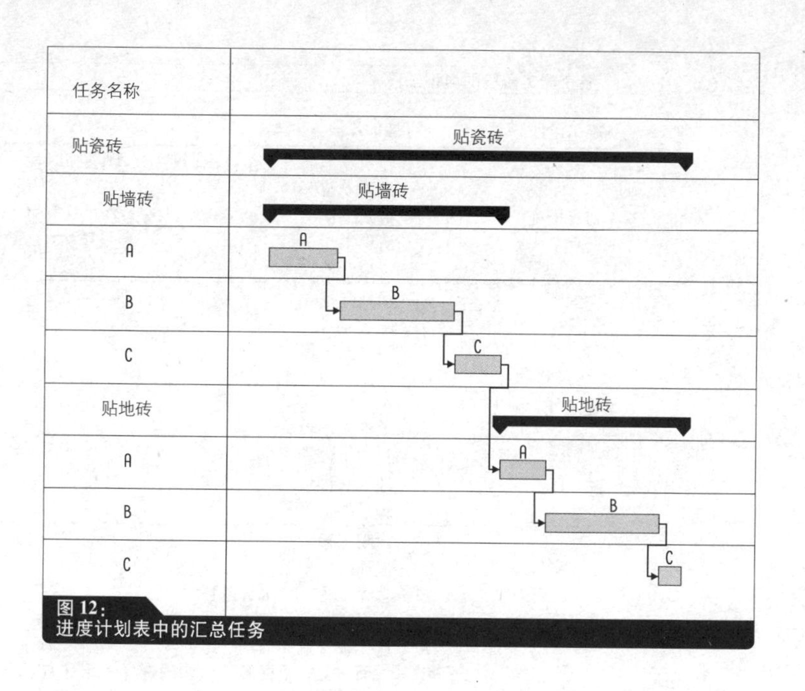

图 12：
进度计划表中的汇总任务

项目进度计划的大致阶段

工作包通常是根据施工进度按照日期顺序进行排列的。若要将整个施工过程进行划分，则可能会包括以下几个阶段：

1. 初步措施；
2. 建筑外壳施工；
3. 围护结构施工；
4. 内部装修；
5. 设备设施安装；
6. 收尾工作。

工段任务安排

通过将不同的施工阶段划分为各自不同的工段（有些情况下也存在工段的重叠），可以形成进度计划的初步纲要。然后，将工作包所包含的一系列任务进行分类。〉参见表 1 以及“设计和施工过程中的工作流程”一章中的内容

表1：
各个施工阶段的典型工段

施工阶段	可能的工段
初步措施	— 建筑场地准备工作（围墙砌筑、现场临时房屋建造、公用设施安装等） — 拆毁工作 — 清理工作 — 场地开挖
建筑外壳施工	— 场地开挖 — 降水 — 混凝土工程 — 砌体工程 — 钢结构工程 — 木结构工程 — 封闭工程 — 地下水排水 — 脚手架工程
围护结构施工	— 封闭工程 — 屋面防水 — 管道工程（雨水管道） — 门窗安装 — 遮光板（遮光剂） — 立面施工（根据维护结构的选择采用抹灰、天然石材、砌块或玻璃幕墙等）
装修	— 砂浆底层 — 找平层 — 干式墙施工 — 金属制品施工（比如，扶手护栏等） — 天然（或者人造）石材施工 — 贴瓷砖 — 拼花地板铺装 — 地板安装 — 墙漆/墙纸
设备设施安装	— 通风系统 — 电力系统 — 卫生设施/配管工程 — 热水系统 — 燃气系统 — 防雷工程 — 运输工具和电梯安装 — 防火工程 — 建筑自动化工程 — 安全设备
收尾工作	— 木工 — 门锁设备 — 完工清理 — 室外设施安装

任务顺序安排和持续时间的确定

接下来的任务是确定不同任务的顺序和计算任务的持续时间。〉参考“进度计划的创建”一章中“任务顺序安排”和“确定任务持续时间”部分内容在此过程中，工程人员不仅需要考虑施工阶段中不同任务之间的依赖性，还需要考虑可能的外部因素的影响。打个比方，工程人员需要注意业主所规定的截止期，而且可能打算在暑假开始之前举行建筑的封顶仪式。另外，在建筑场地周围发生的一些事物也可能会对工程进度计划造成一定的影响（比如，街道或者城市举行庆祝活动、政府部门所规定的设施设备连接日期等）。只要存在可能，均建议工程人员将一些关键的施工任务安排在霜期以外天气温和的时期内。

将工程划分为不同的施工阶段

在进行进度计划纲要编制和确定不同工段施工顺序的时候，最重要的一步是将整个工程划分成为不同的施工阶段。对于诸如楼面砂浆底层铺设之类的任务可能会在若干不同的地方出现（比如地面层楼板砂浆底层铺设、二层楼板砂浆底层铺设等）。将施工阶段进行进一步的划分，能够帮助我们找出不同任务之间的重叠和搭接，而且能够有效地理清施工顺序（打个比方，通过进度计划表可以清楚地发现在所有楼层进行找平层施工之前均需要确保砂浆底层的铺设工作已经完成）。将不同的施工阶段所对应的施工任务标记在进度计划中，能够提醒施工单位明确各自的开工时间和施工顺序。〉参考图 13

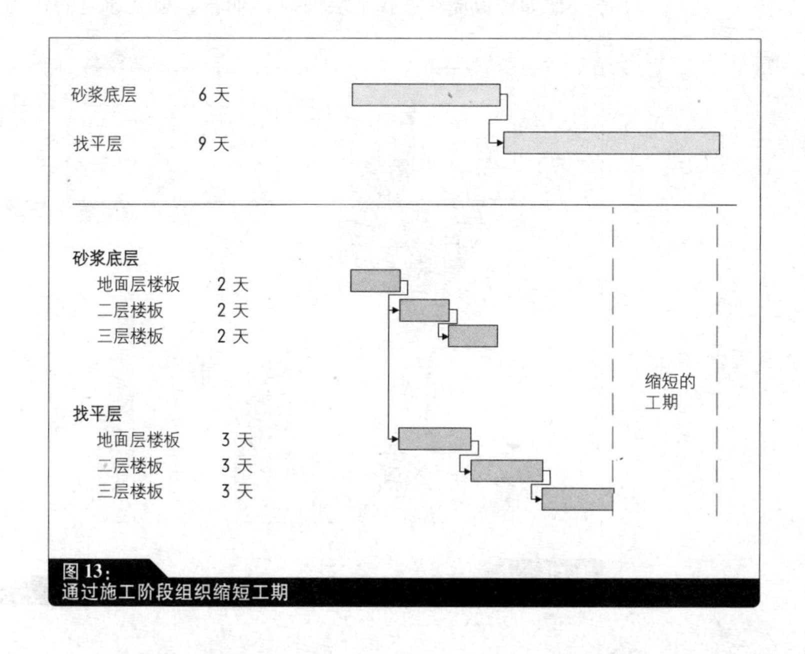

图 13：
通过施工阶段组织缩短工期

工程人员在进行施工阶段划分之前必须进行深入的分析，因为一旦划分之后若要再次进行改变将会导致巨大的额外工作量产生。其经验法则是，将施工阶段所划分成的施工段越小，则施工工期越短。然而，由于受到工程规模和时间的限制，并不会将各个施工阶段进行过于精细的划分，否则在施工进度计划的制订和现场使用时将遇到较大的困难。〉参考“施工进度计划表的使用”一章中“施工进度计划表的更新与调整”部分内容

对于小规模的工程（比如房屋扩建等），可能并不需要进行施工阶段划分。而对于大型工程项目，为了确保在合理的工期内完成，可能需要划分成若干施工阶段。

最后，有时为了使整个工程施工更加合理，物资运输更加方便，也需要进行施工阶段的划分。根据建筑物的外形不同，可以按照建筑物的楼层、楼梯所连接的建筑物不同部分、电梯井各侧所连接的单元以及施工机械的租用顺序进行划分。〉参考图 14

不同的施工阶段有各自需要重点考虑的相关因素，比如独立通道、场地封闭的程度以及不同施工阶段或生产过程中所需的重要建设物资的数量等。

独立通道可能会采用楼梯或者其他的通道形式，对于某些施工任务（比如楼板浇注、找平层铺设）而言，独立通道的作用尤其重要。

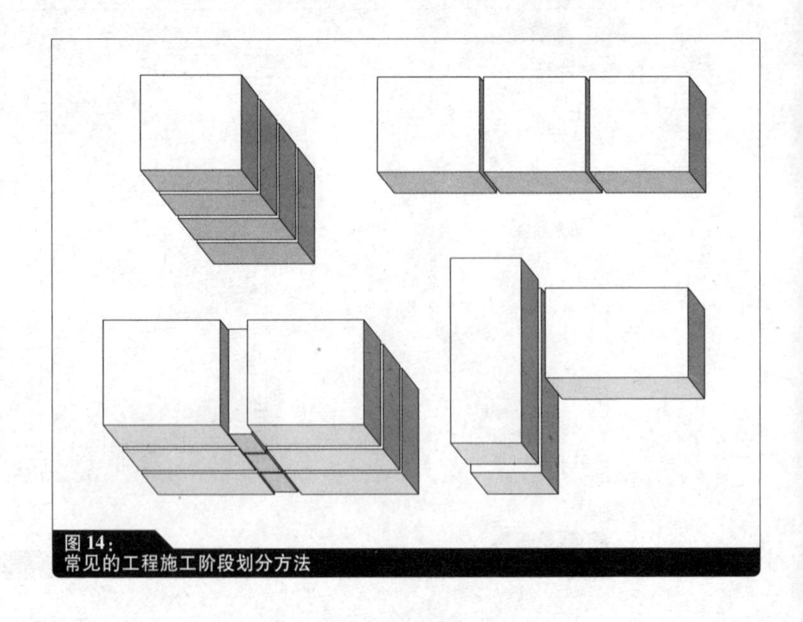

图 14：
常见的工程施工阶段划分方法

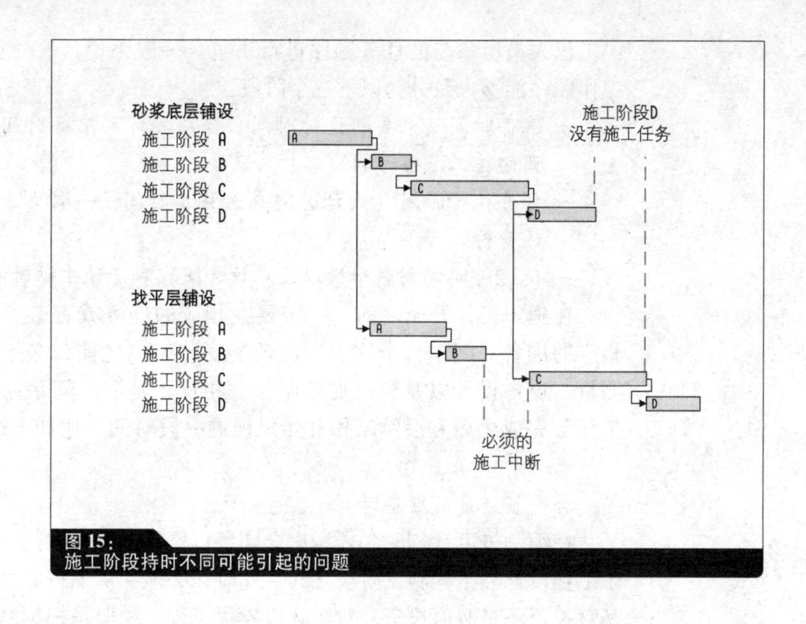

图 15：
施工阶段持时不同可能引起的问题

它能够避免不同的承包商占用其他方的通道，同时也可能防止施工人员在返回驻地时通过封闭区域。

除此之外，场地封闭还可以保护建筑已经建成的部分免受破坏。如果将施工场地进行封闭，仅仅对正在施工的承包商开放，则可以很好地避免建筑物损坏、变脏以及物资失窃等，即使发生这些情况，也更容易确定可能的责任方。

对于工程量的确定，设计者在进行不同施工阶段划分的时候应该尽可能确保不同施工阶段的工程量基本相同，以保证工作循环的连续性，同时避免在不同的承包商之间存在较长的工作中断。〉参考图 15

在施工阶段划分的时候需要考虑的另一个因素是同一个承包商所进行的不同的施工过程。一般来说，建筑结构是一层一层（从下至上）施工的，但是对于一些安装工程施工过程则是沿着安装路径进行的——比如管道安装（从下至上），或者是在施工段内独自循环进行的（比如对一个租赁单元的再划分）。该因素往往是导致不同工作流程相互误解和施工中断的原因。

P31

任务顺序安排

为了对不同的任务有一个更好的认识，下面我们将系统地对某一位工程人员所参与的任务顺序进行讲解。一般来说，他/她所完成的工作顺序能够大致分为以下三个阶段：

— 准备时间（必要的设计时间，合同签订的准备时间以及承包商的施工准备时间）；
— 任务执行时间（设计时间或者施工时间——取决于工程人员的身份）；
— 延迟时间（材料干燥以及养护时间）和后续工作时间。

在施工工作开始之前，一定要在相应的任务或者节点之前安排相应的准备时间。打个比方，在现场安装窗户之前，必须进行尺寸测量、窗户设计以及窗户加工制作。相反，延迟时间是在施工段施工任务完成之后和后续工作开始之前的一段时间，比如干燥时间等。

P32

合同签订前的准备时间

在编制进度计划时必须考虑合同签订所需的准备时间，该时段处于工程设计和工程施工阶段之间。此时，必须弄清工程的性质是属于私营投资还是政府投资。对于政府投资工程，一般是根据基本的方针和法规所规定的工程截止期来签订合同。对于私营投资工程，这些规定则不具有约束力，因此在签订合同时则可以采用不太正式而是更加直接的方式。但即便如此，也应该留有足够的准备时间，以确保所有参与者能够用合理的方式来进行合同的签订。对于私营投资工程，可以将政府工程的相关规定作为参考。

进度计划节点

合同签订的过程包括若干步骤。〉参考图16及“设计和施工过程中的工作流程”一章中“施工准备工作的协调”部分内容进度计划至少需要包括以下几个任务（或者节点）：

— 公示，大多数的政府投资工程必须进行该过程；

提示：

在确定合同签订的准备时间时也需要考虑设计服务中为了确定最有经验、最适合该工作的设计专家所需的时间（比如，在已有建筑的改建过程中需要确定建筑防火方面的专家）。在签订设计合同的时候也可能需要进行公开招标。

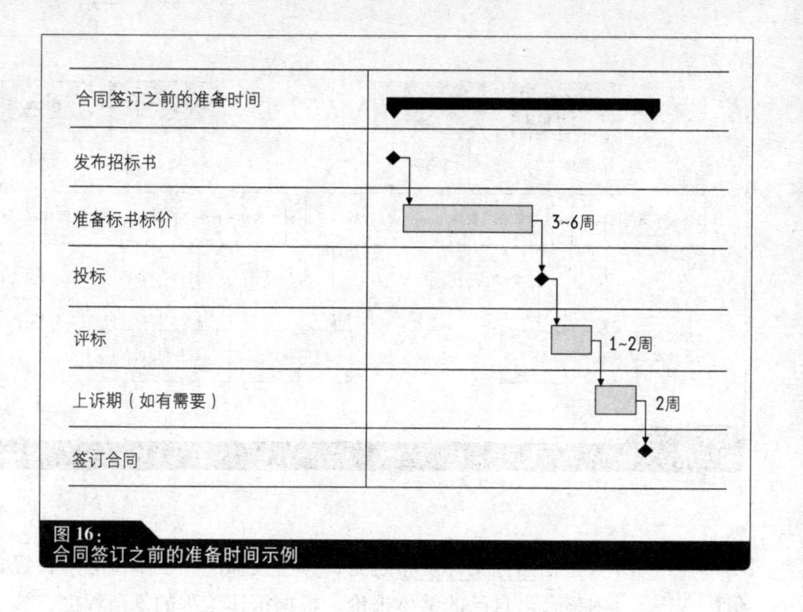

图16：
合同签订之前的准备时间示例

— 发布招标书：在规定的截止期内必须完成所有的文件工作；
— 投标：相关企业必须在截止期内递交标书；
— 签订施工合同：该节点为业主的截止期；
— 工程开工。

由于现场施工一般是按计划准时开始的，所以设计者在为合同签订计划准备时间的时候，应该从开工日期向前推算。

施工合同和开工日期

施工合同的签订时间和开工日期之间应该至少留有两个星期的时间。因为在开工之前，施工企业必须利用这段时间来对施工工作做必要的准备（确定材料的需求量、材料的运输到场以及现场施工设备准备等）。

投标

在投标和施工合同的签订之间也需要留有足够的时间——根据工程的复杂程度，一般至少是1～2个星期。在此期间，设计者将对所有的标书进行评审，并为业主提供相应的报价对比单，业主将在此基础上决定所要雇佣的施工单位。对于标书中所有不清楚和有差错的地方都必须进行讨论澄清。对于政府工程，通常会规定一段时间的上诉期，在此期间未中标的投标人可以申请上诉。对于政府工程，有时决策的过程是比较漫长的，将会超过通常所需的两个星期的时间。

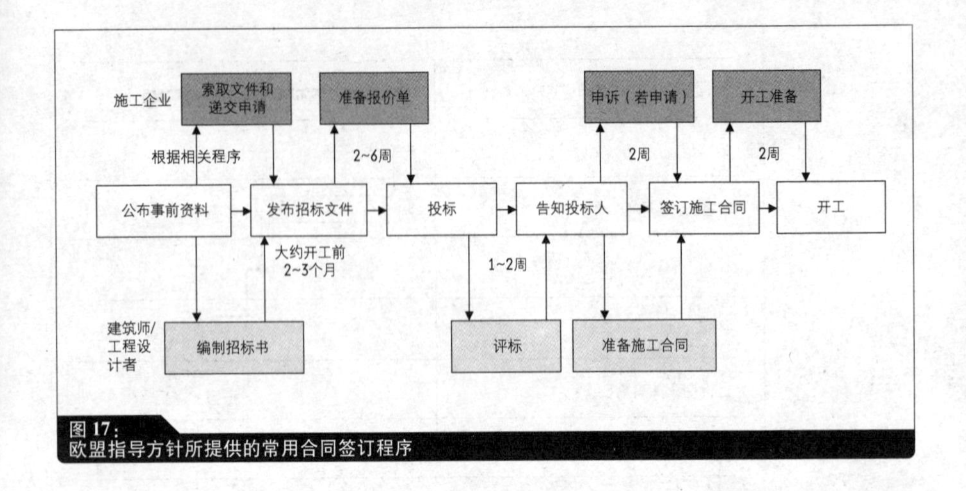

图 17：
欧盟指导方针所提供的常用合同签订程序

发布招标文件

在招标文件发布之后，参与投标的施工单位必须在规定的截止期内整合出自己的整体报价。根据拟建工程的复杂程度，施工单位可能会花费大量的时间和精力来计算自己的报价。因此，必须给施工单位留出足够时间（通常是六周左右）来准备报价。在这段时间内，施工单位需要向材料供应商询价，还有可能进行自身对不同工作段分包商的招标工作。即使没有特别紧急的事情，施工单位一般也无法在两周之内整合出自己的报价。

公示

根据国家法律和所采用的合同签订程序的相关规定，政府机构在进行工程建设招标之前必须提前一段时间进行公示。对于建设信息的提前公示有助于施工单位按时进行文件索取和投标申请。〉参考图 17

P34

承包商的施工准备时间

不是每一个施工项目在合同签订之后就能立即开始施工。对于施工单位而言，在现场施工开始之前，一般还需要进行一些必需的准备工作。设计者在进行进度计划安排时也必须将承包商的施工准备时间考虑在内，对于需要承包商进行计划工作以及存在工地外预制加工和大量材料采购的工程需要特别注意。

计划材料需求量

施工单位在进行材料采购的时候，为了搭建安全的财务框架，一般是在签订采购合同之后直接下订单采购。对于多数的施工任务，比如底层砂浆浇注和找平层铺设——可以采用现成的标准建筑材料——施工单位可以在上述施工合同签订与开工之间的两周内完成采购。

但是对于某些情况，施工单位可能无法在批发市场找到现成的材料。此时，设计者应该提前将计划材料需求量所需的时间考虑到进度计划中。

样品检查

如果业主要求在下订单之前进行样品检查（比如说砖、瓷砖、窗户、涂料以及其他类似材料），〉参考图 18 和图 19 设计者必须为以下步骤预留足够的时间：

— 样品采购；

— 样品检查与认可；

— 样品更换或者其他样品的采购（若有需要）；

— 材料运输时间。

预制

除了材料采购需要时间之外，有些施工项目在现场施工开始之前需要施工单位进行预制构件的设计和制作。

对于某些施工项目，施工单位可能需要提前进行现场测量工作，以确保构件预制和安装的精确性。而测量工作只有在施工过程完成到一定进度之后才能进行（比如说，预制楼板的安装或者建筑外壳的开洞等）。

根据现场测量的结果，施工单位将绘制自己的施工图，该施工图则是进行构件预制的依据。如果合同中规定该施工图必须得到建筑师的认可后方可进行预制构件的加工，那么设计者在进度计划安排中除了考虑施工图的绘制时间之外，还必须留有建筑师检查和认可施工图的时间。〉参考图 20

在得到了建筑师的认可之后，就可以开始预制工作。根据承包商

举例：

对于一些大型施工单位，当合同金额超过一定的限值之后，负责施工工作的相关人员可能对于合同签订没有决定权。必须经过更高级别的机构（比如说企业管理委员会）同意之后才能进行合同签订，该过程取决于该机构召开集中会议的频率，可能需要经过较长的一段时间。

举例：

如果工程中采用的是国外进口的特殊规格天然石材，那么应该首先订制该材料，然后进行加工和运输。如果工程中存在某种材料的需求量特别大，或者某种材料需要进行单件生产，由于批发市场中将出现材料的缺乏，此时材料生产会消耗一段时间。

图 18：
外立面系统检查

图 19：
屋顶边缘设计检查

的技术以及预制材料的性质不同，预制构件从开始制作到可以使用大概需要 6 ~ 8 周或者更多的时间。但是现场安装工作所需的时间则一般相对较短。

常见的需要预制的构件如下：

— 外立面、门窗；

— 玻璃屋顶、天窗；

— 预制混凝土单元；

— 钢结构（比如承载门厅结构、钢楼梯、钢扶手等）；

— 门结构（比如屋顶桁架）；

— 系统构件（比如办公室隔断墙）；

— 通风系统；

— 电梯系统；

— 内置家具、内部设施。

P36

施工工期

施工工期是完成工作包所包含的所有任务的施工时间。设计者在进行任务安排的时候，必须考虑不同任务和承包商之间的依赖关系，同时还要考虑任务的持续时间和不同施工段的任务安排。〉参考“进度计划的创建”一章中“确定任务持续时间”部分内容一般来说，如果某个施工单位与其他施工队伍之间不存在依赖关系，那么在该施工单位所制定的

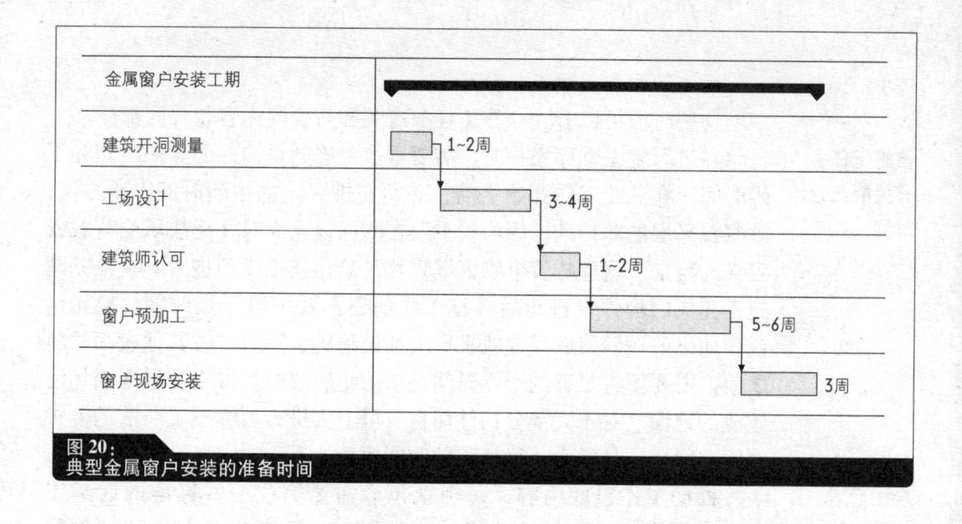

图 20：
典型金属窗户安装的准备时间

进度计划表应该达到详细的深度。对于有些工作区域，可能存在不同承包商交叉施工的情况，此时所制定的进度计划表应该达到非常详细的深度，确保不同承包商能够明确各自的施工时间以及和其他承包商之间的依赖关系。

如果同一承包商的不同任务之间的时间间隔较长，而且任务之间并不存在太强的联系，那么一种有效的办法是将这些任务分成两个不同的工作包来签订施工合同。这样不仅能为不同任务确保足够的准备时间，而且可以确保施工合同的齐备性和一致性。

设计和施工过程中典型的施工任务以及任务的依赖性将在“设计和施工过程中的工作流程”一章中进行介绍。

举例：

在钢结构工程和金属结构工程施工过程中，通常包含一些贯穿整个施工过程的任务，主要有钢结构、窗户、外墙覆层、门、扶手以及楼梯等。由于在大多数情况下这些任务并不需要按顺序进行，而且有些任务可以由专业公司来完成，所以建议将部分任务安排在单独的工作包中。

P38 延迟时间和后续工作时间

这些时段可以分为与施工任务有关和与合同内容相关两部分。

与施工任务有关的延迟时间

在进行施工顺序安排时，需要重点考虑的是构件的养护时间和干燥时间。在后续工作开始之前，必须安排一定的中断时间。比方说，砂浆找平层的养护期，因为找平层在刚铺设完成时并无法承受荷载或者供人行走。这意味着个别区域将暂时禁止施工设备进入。在某一刚施工完成的构件上进行后续施工之前还需要一段干燥时间。换句话说，如果要在砂浆底层或找平层上铺设地砖、涂料或者其他表面材料之前，需要留有足够的干燥时间，以防止后续施工带来的湿度损坏已建成的地面。对于大部分构件而言，可上人进行后续施工所需的养护时间一般比构件完全干燥所需的时间要短。

合同后续工作时间

根据工作包的内容，每个承包商需要完成一些标准的后续工作，这也会对施工工期造成一定的影响。下面举几个后续工作的例子：

— 建筑外壳：在设备安装完成之后，需要封闭所有的墙体留洞；在建筑施工完成之后，拆除现场临时安装部分（如果承包在建筑外壳施工中）；
— 门窗：在建筑施工完成之前，完成门、窗以及把手的安装；
— 抹浆：门洞磨平、楼梯踏步面层以及窗台抹浆；
— 建筑设备：开关、暖气装置以及卫生设备安装；技术系统启用；
— 涂料：瓷砖铺设完成后的涂装工作；施工完成后的墙体、顶棚上的家具安装和固定。

设计者在进行进度计划编制的时候，应该将上述后续工作作为单独的任务安排在进度计划中，以避免施工单位以超过合同工作包的工作时间而进行索赔。另外需要重点关注的是施工的保修期，保修期从竣工验收开始计算。合同规定的施工任务完成越早以及验收工作完成越早，那么保修期会结束得越早，在保修期内业主有权要求对建筑物的损坏进行修复。

当设计发生变更或者在施工过程中发现施工条件的改变时（比如，在已有建筑或者建筑地基中发生一些意外的条件变化），需要进行必要的信息搜集和咨询服务工作，此工作需要的时间属于计划编制方面的后续工作时间。

P39

确定任务持续时间

当计划编制者明确了所有承包商和设计人员的任务之后，下一步是对所有任务的持续时间进行估算。建筑师一般会根据以往项目的经验数值或者通过向承包商、施工单位询问的方式来确定标准任务的持续时间。

另外，可以根据任务数量和与数量相关的时间量来确定任务的持续时间。此时，必须对单位产品生产时间与单位时间生产率两个概念进行区分。

时间定额

时间定额（UPT）是指完成单位产品生产任务所需的工时数量，可以按照下式进行计算：

时间定额 = 消耗的工时数量/任务数量（比如，$0.8h/m^2$）

产量定额

产量定额（UPR）是时间定额的倒数，指的是单位时间内完成的产品数量，可以按照下式进行计算：

产量定额 = 完成的产品数量/单位时间（比如，$1.25m^2/h$）

在建筑行业中，在与施工机械设备相关时一般采用产量定额（打个比方，挖掘机的工作性能一般用 m^3/h 进行表示），而与人工劳动时间相关时一般采用的是单位产品生产时间（打个比方，砌筑 $1m^3$ 砖墙所需要的小时数，一般采用 h/m^3 进行表示）。

确定工程量

工程量估算所采用的数量单位与时间定额、产量定额相同（m、m^2、m^3 或者块等）。如果在土方的产量定额中采用的是 m^3，那么在进行开挖量计算的时候也必须采用 m^3 作单位。

由于在工程量计算时所采用的数量单位与设计过程中的其他阶段（成本计算、投标等）相同，所以工程量可以直接采用其他阶段的计

提示：

时间定额和产量定额通常与施工企业的性质、施工方法以及工人水平有关。另外，现场施工工作通常会受到现场具体条件的影响。所以不可能采用这些近似参数提前计算出精确的任务持续时间。

算结果。若有这些过程中并没有给出工程量，那么就需要从头开始计算。在确定工程量计算精度的时候，设计者需要明白时间定额和产量定额并不是一个精确的数值，所以通常情况下仅仅需要一个大致的计算即可。

任务持续时间的估算

完成一项任务所需要的总小时数被称为"工时"（PH），工时可以根据工程总量与时间定额或产量定额计算得到。将工时数除以工人数（W）和每日工作时间（DWT），就可以得到大致的任务持续时间（D），任务持续时间用工日（WD）表示：

$$D = \frac{PH(UPT \times \text{工程总量})}{W \times DWT} = [WD]$$

工人数量

工人每日的工作时间通常是根据工资支付规定的相关条文来确定的，只有在诸如工期压力非常大等特殊情况下才允许进行加班。在进行全面考虑之后，应该对工人数量进行优化考虑，以确保施工过程有效地进行。对于有些任务（比如窗户安装等），存在所需的最少工作人数——否则将无法有效完成或者根本无法完成任务。但是，工人数并不能随意增加，因为如果工人数太多，在随后工作中有可能无法进行有效的人员部署。以砂浆铺设为例，所需的工人数量很大程度上取决于可以使用的机械数量，工人数量增加对工作效率的提高非常有限。

对人员配备进行计划安排是施工企业内部为了确保合理工期的一种手段。通常，计划编制者让施工单位在给定工期的基础上进行工人的雇佣工作。即使如此，项目经理通过人员数量的计算可以确定工地是否会处于工人缺少的状态，避免由于人员缺少而影响最后的工期。

注释：

由于工地条件的差异，计算得到的任务持续时间可能会和实际时间相差达到50%，所以计算中所存在的细小误差完全可以忽略。如果需要进行精确计算，那么应该对不同的时间定额来源进行比较分析。附件中列举了大量常用的时间定额，可供计算查用。

举例：

如果时间定额是0.8h/m^2，工程量是300m^2，工人总数为5人，每天工作时间8小时，那么所需任务持续时间为：

$$D = \frac{0.8 \times 300}{5 \times 8} = 6WD$$

根据任务持续时间的计算公式进行反算可以得到完成某项工作所需的工人数量。

$$W=\frac{PH(UPT\times 工程总量)}{D\times DWT}$$

前文所介绍的计算结果也可以在合同签订过程中作为对施工单位现场工作能力的一项评价标准。

若所需的工作人数是根据不同施工阶段的任务持续时间来专门确定的，那么将为进度计划编制者带来非常大的便利。假设不同承包商将依次进入某一施工阶段进行施工，然后再依次转到下一施工阶段，如果不同承包商的工作持续时间相同，那么可以确保各个不同施工阶段的工作连续进行，而且不同任务的施工人员不会存在停工等待的时间。〉参考图 21

任务持续时间确定结果

通常情况下，并不需准确计算每一个单独的计划表和每一项单独的任务的持续时间，所需要只是在经验数值的基础之上对持续时间进行一个估算。其主要原因是因为所有的施工项目中的任务持续时间都会经过一些小的改动，而且大多数情况下起到限制作用的是整个建筑的完工截止期。对于整体建筑完工截止期而言，更重要的是施工任务的顺序安排，这部分内容将在下一章内容中进行阐述。如果施工顺序发生错误，那么将可能导致结构发生变形或者更加严重的后果。

然而，也不能忽视任务持续时间计算的作用，因为它是与施工单位签订合同时计算工期的基础。所以，在实际工程中应该采用的是符合工程实际的任务持续时间，以确保不同施工合同中的任务能够顺利地实施完成。

举例：

假设完成某一项工作所需 240 工时，而根据时间计划表，该任务需要在 5 个工作日内完成，那么所需的工人数应该如下计算：240*PH*/（5 × 8）（*PH*/*D* × *DWT*）。故所需的工人总数为 6 人。

注释：

进度计划编制者在确定任务的持续时间时，不仅仅需要采用数学计算的方法，还要考虑其他相关因素，以确保施工工作在全年的时间内（比如假期、寒暑期以及霜冻期等）以相同的速率进行，包括圣诞节前后和新年前夕——尽管可用工作日的时间就能满足工期的要求。

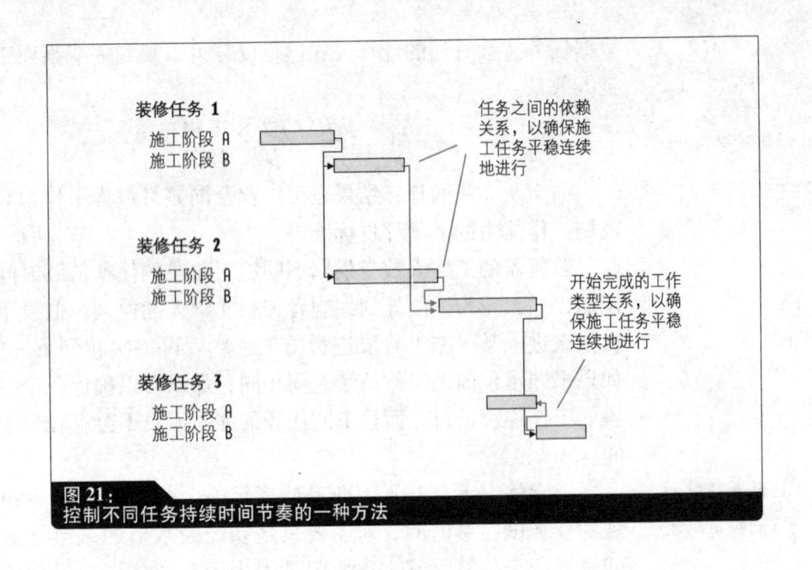

图 21：
控制不同任务持续时间节奏的一种方法

任务持续时间确定结果

一般来说，建筑设计工作的持续时间无法用时间定额来进行计算，因为脑力劳动和创造性工作是无法用单位产量所需的工作时间来进行衡量的。设计工作持续时间的确定和项目规划阶段类似，是经过设计者和相关专家的交流讨论之后确定的。这种方式能够充分利用设计者的经验和工作时间。通过向设计者说明设计阶段滞后将可能带来的后果之后，能够使设计参与人员明白自己的工作能为整个工程按时完工投入使用所起到的作用。

P43

设计和施工过程中的工作流程

本章将介绍设计过程和施工过程中的工程人员所需完成的标准任务，并对不同任务之间的相互依赖关系进行阐述。本章的学习目标是能够运用这些背景知识对实际工程进度计划表中的相关施工任务进行辨别和描述。

P43

设计阶段参与人员

在工程设计阶段，需要进行相互协调的不同参与人员大致可以分为以下几类：〉参考图 22

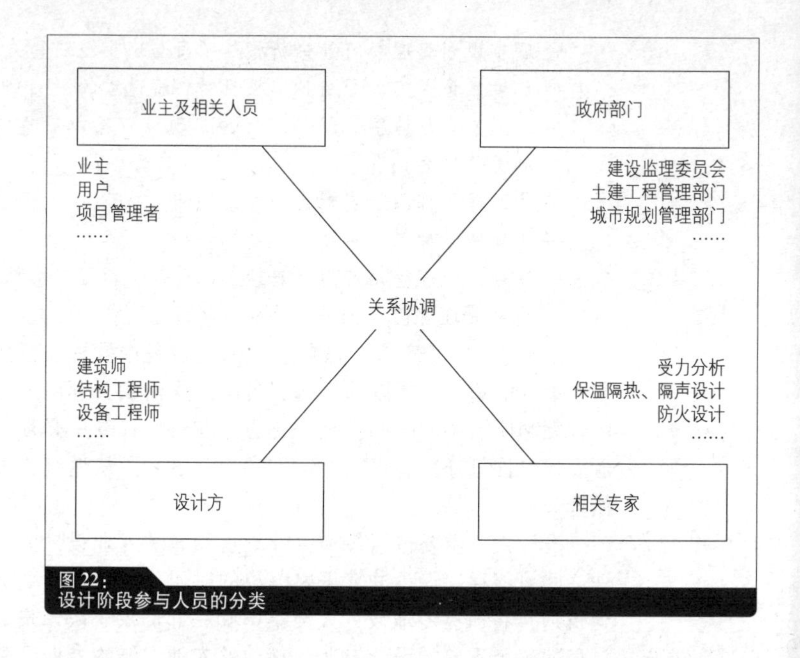

图 22：
设计阶段参与人员的分类

业主及相关人员

作为工程项目的发起者，首先要提到的是项目的业主或者业主委托方。业主可能是一个单独的人，也可以是多个人、机构的综合体。业主性质的不同使其与项目开发商、财政资助方（银行）以及将来的用户对项目的看法大不相同。打个比方，假设某项目的业主是一个公司或者某事业单位，那么项目经理必须向对该项目具有决策权和影响力的相关委员会或者部门负责，而且必须确保这些部门或者委员会参与决策制定的过程中。

对于计划编制者，只有弄清建设方的决策过程以及所需的时间，才能更好地协调建设方公司或者组织内的参与人员，使之在给定的时间内做出决策。

政府部门

每一栋房屋的新建都需要和相关的政府部门进行若干的交互过程。这些政府部门将决定新建房屋的合法性，并根据相关法律颁发建筑许可以及对建筑的使用过程进行监督。而这些政府部门在此过程中所参与的程度则很大程度上取决于拟建建筑的类型、功能以及相关的法律条文和当地条件。除建设监理委员会以外，还可能涉及以下一些相关部门：

— 土建工程管理部门（拟建项目与公共基础建设项目有关）；

— 城市规划管理部门（分析拟建项目与城市规划的关系）；
— 环境保护部门（分析拟建项目对环境造成的影响）；
— 职业安全与健康管理部门（施工现场以及以后建筑内部工人的工作安全）；
— 古迹保护部门（针对古迹建筑）；
— 土地测量部门（地图，现场测量）；
— 房地产管理登记部门（房地产的管理、留置以及限制约束）；
— 贸易管理部门（如果房产随后将用于商业用途）。

由于这些政府部门一般将会在建设过程中发挥监督作用或者扮演决策者的角色，所以计划编制者必须弄清相关部门进行决策过程的步骤和所需要的时间。打个比方，在进度计划中必须为相关部门颁发建设许可证留有可行的时间段——该时段从递交相关文件后开始算起。

计划员

设计人员主要包括各项目计划员和相关专业设计人。当不同专业之间可能存在设计冲突需要解决的时候，由项目计划员（通常是建筑师）将所有专业设计人员集中到一起商议解决。贯穿整个设计过程的三个重要的设计专业包括建筑专业、结构专业和建筑设备专业。但在实际工程设计中，需要有大量不同专业的设计人员参与其中：

— 结构工程；
— 室内建筑；
— 电气工程；
— 给排水工程；
— 通风系统；

举例：

假设在某项目的增补设计过程中，建筑师需要业主决定楼板覆层或者其他表面的建筑材料。那么建筑师需要在项目的早期阶段向业主提供相关的样品以及相应的替代材料，并告知不同材料的优缺点（比如，成本、寿命、材料敏感性等）。业主可能需要和相关人员或者将来的承租者对不同的材料选择进行讨论。

— 防火工程；
— 数据处理；
— 厨房设计；
— 立面工程；
— 景观和开放空间设计；
— 照明系统；
— 设备管理。

设计专家

除各专业设计人员外，还需要相关的设计专家对不同领域的设计工作进行评估并提交审查报告。设计专家必须至少对设计的隔热、隔声、防火和受力进行评估审查。

在编制进度计划的时候必须将专家的评估审查过程考虑在其中，特别是审查报告的提交。打个比方，在获得施工许可证开工之前必须提交专家的审查意见。所以，在雇用相关专家进行设计审查的时候，必须留有适当的准备时间。

P46

设计协调

设计阶段专业协调的频率很大程度上取决于建筑结构的规模和复杂性，或者是业主所规定的频率。对于住宅建筑，设计工作的很大一部分专门由建筑师来承担，整个设计过程只和少数工作的截止期相关，比如建筑许可和开工证的申请与颁发等。然而，对于大型项目，比如图书馆或者专业生产设施等，通常需要一系列的专业设计人员参与其中。

设计阶段的划分一定程度上仅仅取决于建筑设计的顺序，因为其他专业的设计人员一般是分别进行各自专业的设计。组织设计阶段最好的方式是在设计的最初阶段建立三个重要专业之间的联系，包括建筑专业、结构设计专业和建筑设备专业。其原因是因为这些专业的设

提示：

关于工程设计和施工过程中的参与人员以及相互工作关系的介绍，可以参考本套丛书中的《工程项目规划》一书。（中国建筑工业出版社，2010 年出版，征订号：19521）

计将对整个设计过程具有持续深远的影响。在设计过程中，各专业设计人员需要提前了解项目的整体规划后才能进行各自专业的设计。这种网络工作方式可以按照以下顺序进行：

1. 项目计划员对工程项目进行前期开发，得到各专业设计人员的设计条件；
2. 向各专业设计人员提供设计条件；
3. 各专业设计人员进行专业设计；
4. 各专业设计人员提交设计结果；
5. 项目计划员对各专业设计结果进行整合和协调；
6. 各专业设计人员进行相互协调并进行再版设计（如果需要）。

在进行进度计划编制时，应该在各专业设计人员提交设计结果之后留有足够的时间，用于协调各专业设计和项目规划设计或者其他专业设计领域之间所存在的冲突。〉参考图 23

整合其他参与人员

整合其他参与人员相对会简单些，因为他们的工作不像上文提到的计划员那样连接紧密。通常会在进度计划表中加进一段时间，以便设计专家能应对整体计划过程中的特殊情况。这就能解释清楚不同的参与人员需要别人提供他们在项目中的专业文献信息，结果也会被整合进整体计划过程中。

设计工作节点

在进行进度计划编制时，最重要的一项工作是确定设计工作的节点，以确保设计成果能够传递到其他不同的工程参与人员（包括业主、设计人员以及设计专家等）手中，并为他们的工作提供基础条件。〉参考“进度计划的创建”一章中“进度计划的基本要素”部分内容通过确定所有参与人员一致同意的时间进度，能够促使所有参与人员严格遵守相关的截止期，防止由于设计人员的耽搁而造成整个设计过程的严重延迟。

注释：

附录中介绍了建筑工程师、结构工程师以及建筑设备工程师在不同工程阶段进行信息交流的常用方式。但是，具体到某一个工程，根据建筑功能、设计以及参与人员的不同，这些方式在细节上将会存在差异。

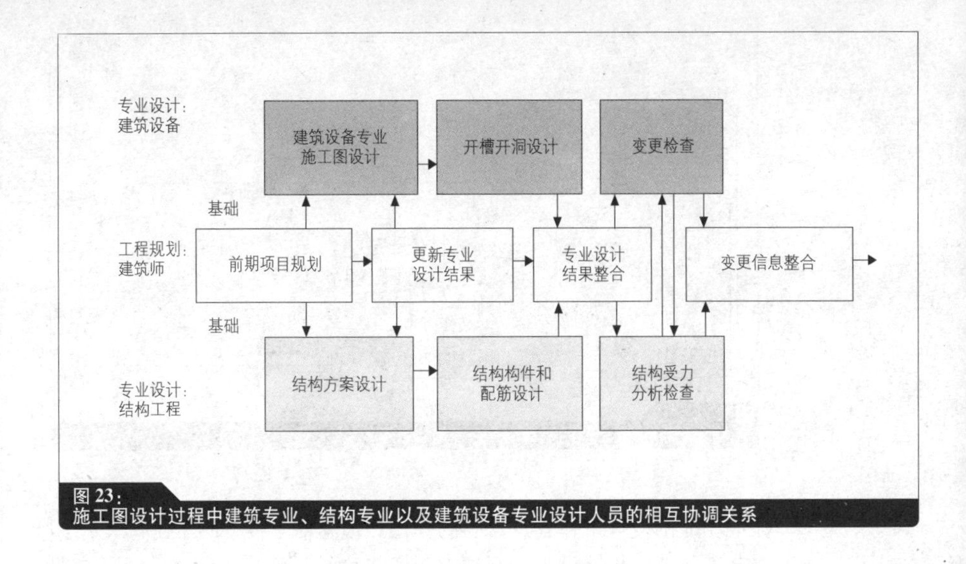

图 23：
施工图设计过程中建筑专业、结构专业以及建筑设备专业设计人员的相互协调关系

P47

施工准备工作的协调

在签订施工合同之前，必须先确定不同工作任务的截止期。施工准备工作是一个长期的过程，通常需要数月的时间。如果已经确定了施工单位的确切的开工时间，那么可以在开工时间的基础之上制订施工准备工作各阶段的进度计划表。〉参考“进度计划的创建”一章中“任务顺序安排”部分内容如果业主是公共机构；那么法定截止期意味着只有在少数几种情况下才允许设计工作发生延迟，否则将直接影响项目的开工时间。

通常一个工程项目中会涉及一系列不同的施工单位，所以在施工准备过程中需要限定相应一系列不同的截止期。因此，比较有效的一种方法是对各个不同施工单位的准备过程分别进行组织计划，以确保所有的前期工作（比如，获得业主的确定方案、施工过程设计、组织和公示招投标工作以及签订施工合同等）能够按时启动。

在进行进度计划表编制的时候同样还要考虑由于施工工作流程所导致的依赖关系。

施工过程中的建筑设计

在许多进度计划概略表中都将建筑设计和施工工作作为两个独立的阶段对待。但在实际工程中，施工图的设计和工程施工工作很大程度上是平行进行的。其原因在于通常施工工期较紧，所以在施工工作开始的时候设计工作并不需要完全结束。虽然说在施工工作开始之前

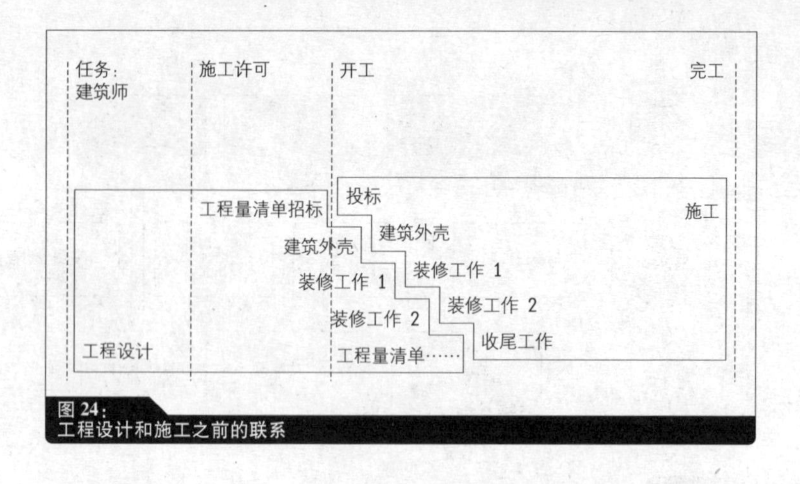

图 24:
工程设计和施工之前的联系

完成相应的设计工作具有重要的作用，但是对于许多施工工作（比如涂装工作、地板铺设等）只要在施工进度满足施工条件之后即可立即开始，也就是说相应的设计工作和合同文件工作可以在施工开始之后再进行。〉参考图 24

在进行施工过程中的平行设计时，会一直存在一定的风险——随后进行的详细施工图设计细节可能会和已经施工完成的部分存在一定的差异。许多建筑构件会在受力、建筑设备、结构体系以及外观方面对建筑的其他部分产生相互影响。

由于一些单独的建筑构件（比如窗户、门以及干作业施工构件等）与其他工段的联系性较强，所以对于这些构件的设计工作需要提前足够的时间，以避免施工后的返工修改。〉参考图 25

P49

施工过程准备

在施工过程中，不同的施工单位分别与业主签订施工合同，而项

举例：

当建筑外壳设计完成之后，下一步需要考虑的是建筑的立面节点做法、楼板与楼梯井高度和混凝土墙体表面的处理方法，以及相对应的施工要求等。另外，在设计阶段的最早期就应该明确位于建筑底板以下下水道系统的设计。

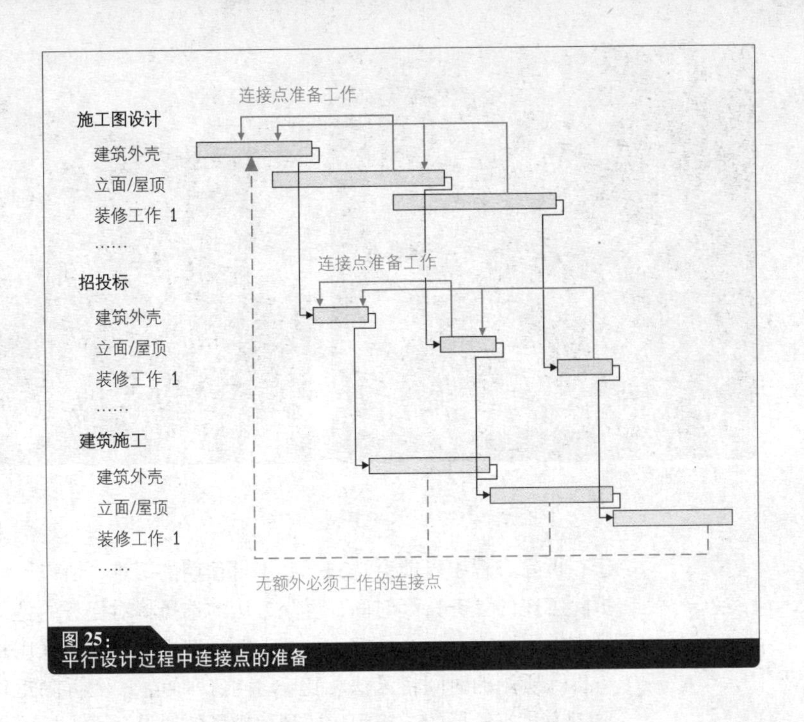

图 25：
平行设计过程中连接点的准备

目规划师和项目经理需要协调所有不同施工单位之间的关系，其目标是确保施工工作无障碍地协调进行。协调工作的主要任务是处理不同任务和承包商之间的连接点，常见的连接点将在下文中加以描述。需要注意的是不同工程项目中的连接点可能会截然不同。

在建筑外壳施工之前，需要对施工场地进行一系列的准备工作。首先，确保施工场地满足施工工作所需的条件。主要工作包括：移除已有场地植被，夯实地基；将已有管道系统进行定位和保护；拆除已有建筑物或构筑物（墙体、围墙等）。

施工场地准备

施工准备工作包括施工场地的准备。这项工作主要包括用于施工管理的现场活动房的安装、联系设备工具的设置以及围墙的砌筑（用于阻止非法进入）。进一步的工作可能包括通道条件的改善（比如修建出入道路），以及将场地和外部环境进行围护（比如对场地进行遮挡和隔声等）。

拆除工作

对于一些新建房屋，场地准备工作一般可以在几天或者几周内完成。但是，有些工程项目可能会涉及对原有建筑的拆除工作。这种项目在进度计划编制中，施工准备阶段的时间将会显著增加。同时，由

图 26：
已有建筑的人工拆除

图 27：
已有建筑的整体机械拆除

于在拆除过程中可能会遇到一些不可预期的困难，所以一般很难估计拆除工作的任务持续时间。拆除方法的选择将对任务的持续时间产生决定性的影响，打个比方，采用人工轻型机械拆除和采用重型机械进行拆除所用的时间将无法相比。〉参考图 26 和图 27 在拆除工作时，还需要考虑建筑垃圾的运输和运送到垃圾场的通道。因为拆除工作先于施工工作，拆除工作的延迟将直接影响后续工作的进行。

P51

建筑外壳

建筑外壳的施工包括一系列的不同施工任务，这些任务的目的是建立起结构的基本框架。对于混凝土建筑，这些任务主要包括：

— 基坑开挖；
— 砌体工程；
— 混凝土浇筑；
— 脚手架安装；
— 建筑封闭，以防止地面或墙体潮气以及地下水对建筑造成破坏；
— 单独的屋顶建造（如果需要）。

根据建筑类型的不同，在建造过程中可能还需要进行钢结构或者木结构的施工。一般来说，建筑外壳的施工任务由承包商来组织。对于建筑师而言，其主要关注的是建筑外壳施工完成之后所进行的各工段工作之间的连接点。

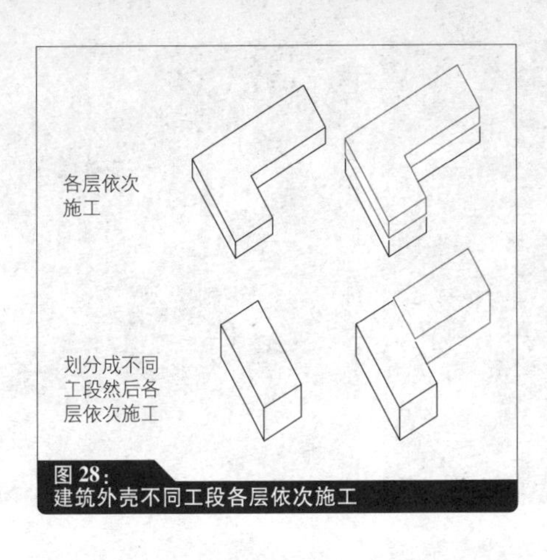

图 28：
建筑外壳不同工段各层依次施工

建筑外壳施工所涉及的任务顺序通常是层次清晰而且意义明确的。当建筑的基础和下水道系统施工完成之后，将依次进行各层的施工。但对于含有大面积楼板的工程，可能会在竖向工段划分之后进行各层的依次施工。〉参考图 28

建筑外壳施工中的预制工作

对于由预制混凝土构件、钢结构构件或者木结构所组成的建筑，其预制构件一般是在场地外预制而成的，然后运输到施工现场进行快速安装。〉参考图 29 以及"进度计划的创建"一章中"任务顺序安排"部分内容除了在楼板结构中经常采用不同类型的预制材料以外，在屋顶结构中也经常采用预制钢结构或者木结构。预制结构的制作安装通常与建筑外壳的施工分别签订承包合同，在制定任务进度计划时需要将该因素考虑在其中。

P52

建筑封闭

围护结构施工

当建筑外壳（包括屋顶结构）或者某一工段施工完成之后，需要立即对建筑与外部环境进行围护封闭。围护结构的施工是进行所有后续装修工作的前提条件，因此在建筑外壳施工完成之后其围护封闭工作越快越好。围护结构需要满足以下功能要求：

— 防雨功能（保护装修材料，确保建筑外壳干燥）；

— 防风功能（特别是在冬天，保存建筑室内的热量）；

— 安全防备功能（防止装修材料失窃丢失）；

— 加热功能（只有在冬天会用到）。

图 29：
预制混凝土墙体的施工

图 30：
混凝土地板和支护的现场浇筑

门窗

达到“建筑封闭”状态的前提条件是对建筑开洞和屋顶的封闭。可以在建筑外壳施工完成后立即安装门窗来达到封闭的状态，或者通过砌筑、安装防盗门的方式进行临时封闭。对于有些建筑类型，可以在“建筑封闭”之后立即进行保温隔热工作以及外墙立面施工。对于局部进行加厚外墙设计的地方，可能需要对脚手架的设计进行修改或者缩短。

屋顶和天沟的封闭

坡屋顶的覆盖和平屋顶的封闭是进行建筑封闭的必要任务。对于天窗、采光穹顶以及管道系统和天沟，同样需要进行封闭。对于设有天沟的平屋顶，施工人员必须确保建筑封闭之后，雨水能够有效地排放到建筑外。

防雷保护

在进行立面和屋顶施工的脚手架拆除之前，必须完成建筑的防雷保护工作，而且避雷针必须确保接地。

P53

装修工作

装修任务的协调是施工监理过程中最需要注意的工作部分。此时，制定精确的任务进度计划安排非常重要，因为不同任务之间的关系紧密，而且不同任务的施工人员可能会分属不同的施工单位——与建筑外壳及围护结构的施工完全不同。

由于任务之间的联系错综复杂，大多数承包商无法协调自己的施工任务与其他承包商施工任务之间的关系。所以，在进度计划表中应该详细地标明任务之间的相互依赖性。

图 31：
连续框格窗的安装

图 32：
阁楼构件的安装

通常情况下，在建筑封闭之后立即进行的是抹灰。由于在大多数情况下，隔墙的表面需要进行抹灰遮盖，所以在抹灰之前需要确保隔墙砌筑到位。对于这种情况，要避免忽视门的传动装置、防火设施以及应急照明系统的设置。对于工业建筑，管线通常布置在墙体外侧，这就意味着在设备安装工作开始之前可以完成抹灰工作。

抹灰

对于抹灰工作而言，门框的安装工作是最常见的一个连接点。根据门框类型的不同，可能会在抹灰工作之前或者抹灰工作之后进行门框安装。打个比方，对于钢结构转角门框需要在抹灰之前进行安装，否则随后对钢门框和内弧面的遮盖会带来额外的成本。而对于双截面闭合门框则必须在施工过程的后期进行安装，以避免在施工过程中被损坏。只要在建筑构件存在相互交叉的地方，就存在连接点，进而会对施工顺序造成影响。比如窗户的安装和窗台的施工、楼梯踏步的施工和楼梯扶手的安装等。〉参考图 33 以及图 37

为了确保建筑表面满足后续施工的要求（比如，涂装工作），在进度计划表中根据抹灰层的厚度应该留有相应的干燥时间。

另外，建议在进度计划表中安排一段补抹灰时间，用来修补在随后施工过程中对建筑表面造成损坏的地方。

找平层施工

由于在找平层铺设或者养护期间无法进行其他任何任务的施工，所以在进度计划表中一般对找平层施工进行专门的规定。在进度计划表中不仅要规定找平层的铺设时间，同时还需要根据找平层所采用的材料留有相应的养护时间。对于水泥砂浆找平层，通常的养护时间将达到 3 ~ 10 天（取决于配合比、天气以及找平层的厚度），但水泥砂

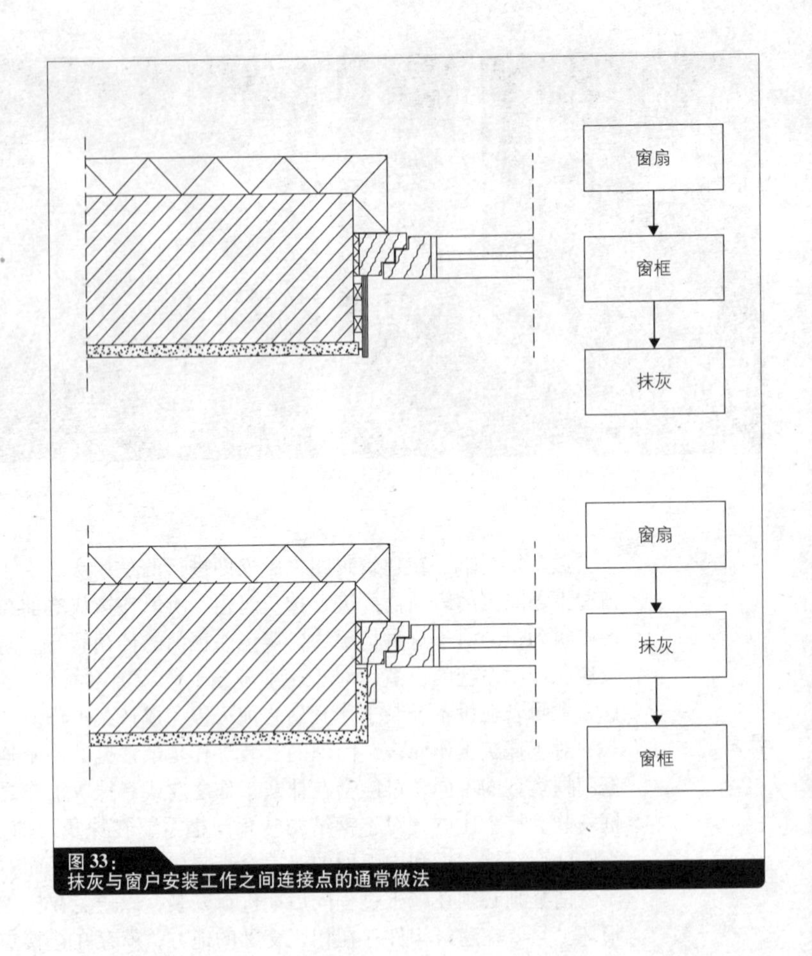

图 33：
抹灰与窗户安装工作之间连接点的通常做法

浆找平层与其他材料相比（比如沥青掺材料，即使沥青材料可以立即施工而且养护期只要两天）具有显著的经济效益。

除了养护时间之外，另外一个重要的问题是找平层的干燥时间。因为地板的铺设只有在找平层有效干燥、湿气足够低之后方可进行。找平层的干燥时间与找平层的材料、厚度以及环境因素（温度、湿度等）有关。在很多情况下，通常将地板的铺设工作放到施工过程的后期，以避免湿度对地板的损坏。如果工期要求比较紧，可以通过使用添加剂或者干燥设备来缩短找平层的干燥时间，但此时会相应增加施工成本。

在进行找平层施工进度计划安排的时候，编制者还需要考虑拟铺设在找平层内的管线施工。比如：地采暖系统、散热系统、地板电源插座以及电缆线路等。〉参考图 34 由于在找平层刚刚铺设的时候无法进

图 34：
增厚找平层铺设之前的设备安装

图 35：
吊顶封闭之前的设备初步安装

入工作面中，所以要检查找平层的施工是否会对其他相关的工作带来阻碍，包括人员通道（以及疏散通道）、材料运输通道以及相关的安装区域（比如电缆安装）等。

干式墙施工

对于干式墙和找平层的施工顺序安排，主要取决于是优先考虑隔声效果（在铺设找平层之间进行墙体施工）还是优先考虑墙体的灵活性（在铺设找平层之后进行墙体施工）。如果一栋建筑中很大部分的墙体和顶棚都采用了干式墙的施工方法，那么在进度计划表的编制过程中，协调干式墙的施工与其他工作之间的关系将是一项非常重要的工作。

由于上述工作之间的依赖关系，石膏板墙体的安装一般分为两步进行。首先，安装墙体的支撑结构，并封闭单面墙体。这项工作由建筑设备安装（包括电力、卫生设备、加热设备以及通风设备等）承包商来完成，而墙体承包商只需要在所有设备安装工作完成之后将墙体的另外一面封闭即可。

对于顶棚吊顶工作，同样存在设备安装工作和干式墙安装工作的协调。在吊顶安装之前必须确保所有的设备安装工作已经完成，而且设计人员必须充分考虑设备、吊顶以及吊顶面板之间的空间几何关系。〉参考图 35

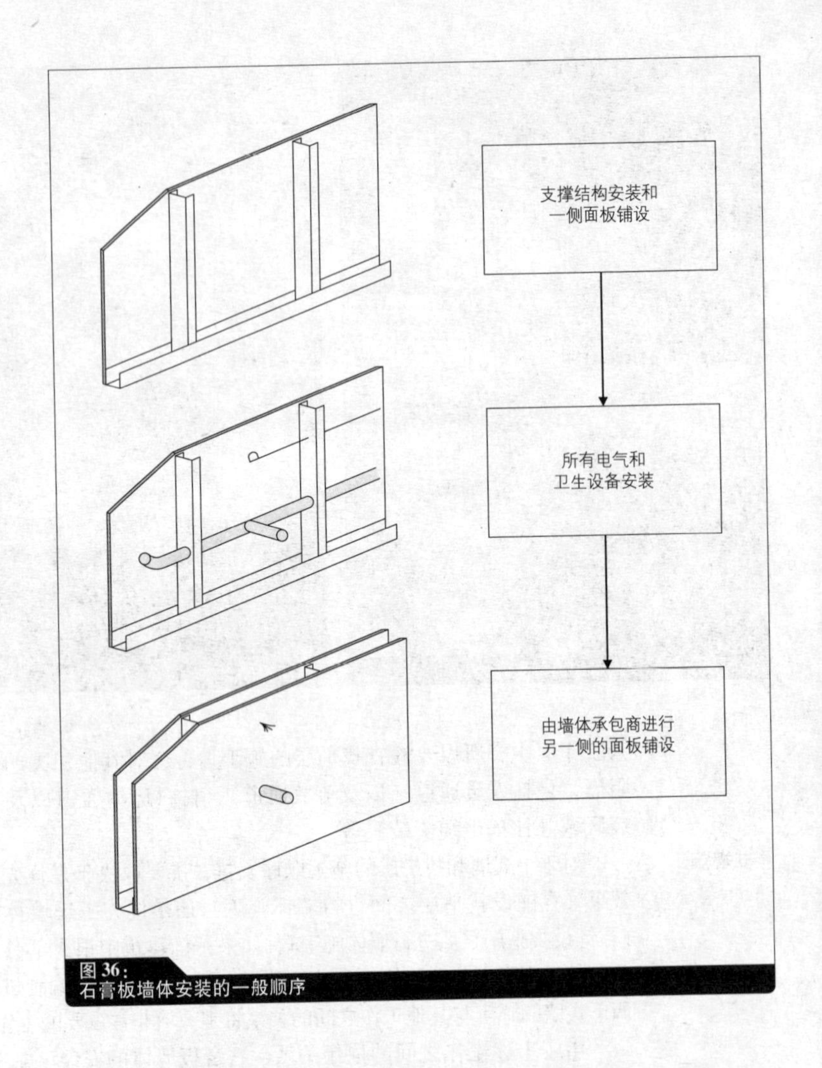

图 36：
石膏板墙体安装的一般顺序

另外，对于一些建筑设备构件的安装需要对干式墙的表面进行一些特殊的准备和处理工作（比如，嵌入式灯具、检修口以及火警探测器等）。〉参考图 36

门与隔墙的安装

对于石膏板隔墙，门框的安装应该和隔墙的安装同时进行，以确保二者紧密连接而且侧面平齐。对于实心墙，门框则是在抹灰工作开始之前或者完成之后采用转角、平面或者槽型截面框架的形式进行安装。〉参考图 37

除以上因素之外，影响门安装方式（包括安装于建筑外壳和干式

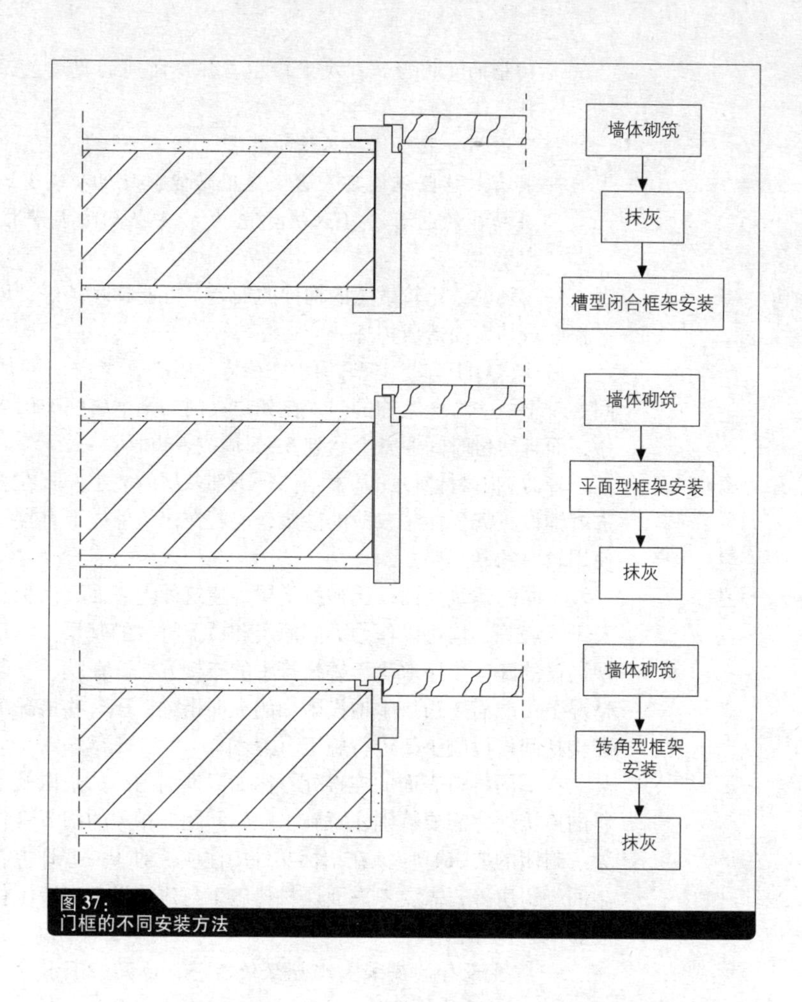

图 37：
门框的不同安装方法

墙中的门）和开始时间的另一个主要因素是门框以及门的类型。对于普通门，门框一般在抹灰开始之前、完成之后或者在干式墙的施工过程中（根据实际施工情况）进行安装，而门扇的安装则是越晚越好，以避免随后施工可能带来的损坏。金属门、标准钢窗以及相应的门框、窗框、配件、面板等一般是作为成品整体进行安装。

在很多情况下，决定门的安装时间的是一些非常细节的因素，比如：

— 门框是否会安装在地面的凹槽中（取决于找平层的施工）；

— 门扇是否与门框同时安装（取决于找平层的施工）；

— 门框的几何形状：决定门框与抹灰之间的覆盖关系（取决于抹灰工作）；〉参考图 37

— 防火门安装对抹灰工作的要求；

— 具有入口监视设备、逃生通道功能、残疾人通道或者自动开关功能的电动门（取决于电力系统以及防火警报系统的安装）。

一般来说，比较敏感的构件的安装尽可能越晚越好，以避免施工过程中对其表面造成损坏。

另外，对于有些类型的门的安装，还必须考虑门的加工交付时间。对于一些特殊结构的门，比如防火门、全金属框架门需要进行定做，而且交付时间通常会达到 6 ~ 8 周或者更长。

贴瓷砖以及拼花地板、石材、地板铺设

瓷砖和石材铺设的基本条件是下部基层的完工。若在平整基层上进行铺设只需要使用较薄的砂浆，但若在建筑外壳上直接铺设则需要使用较厚的砂浆层。

不同的建筑表面，比如找平层、建筑外壳表面、面板、砂浆层以及干式墙等，均可以作为不同面层和覆盖材料的基层。对于楼梯，面层铺设的施工顺序取决于楼梯扶手的连接方式。在扶手安装中，有些情况下可能需要为扶手增设附加的纵向桁梁，还有些情况下可能需要在楼梯间搭设脚手架进行施工。〉参考图 38

在不同材料表面的连接处，比如门框周围的抹灰以及不同地板材料的连接处，需要特别注意施工顺序问题。相关的细节（比如对于支架、封闭的要求）应该在招标文件中注明。对于一些地方，特别是卫生间、厨房等，需要考虑面层材料施工与建筑设备安装工作之间的相互影响，比如：

— 卫生设备的安装：比如坐便器芯、地漏、下水管道、水管配件以及检修孔等；

— 加热设备的安装：热水管、加热元件等；

— 电气设备的安装：开关、地面插座等。

此处需要提醒的一点是在进度计划编制中经常会忽略一些特殊表面材料的施工（比如电梯内地板的铺设、厨房后挡水的贴砖等）。

只要有可能，地面材料的安装均应该以受到损害的风险最小的顺序进行安装。打个比方，地毯、铺地板塑料或油布应该在尽可能晚的阶段进行安装，因为它们比拼花地板、木地板以及石材更容易弄脏和损坏，同时它们的铺设速度也非常快。这些地面材料的铺设一般是作为施工过程的最后一道工序进行。

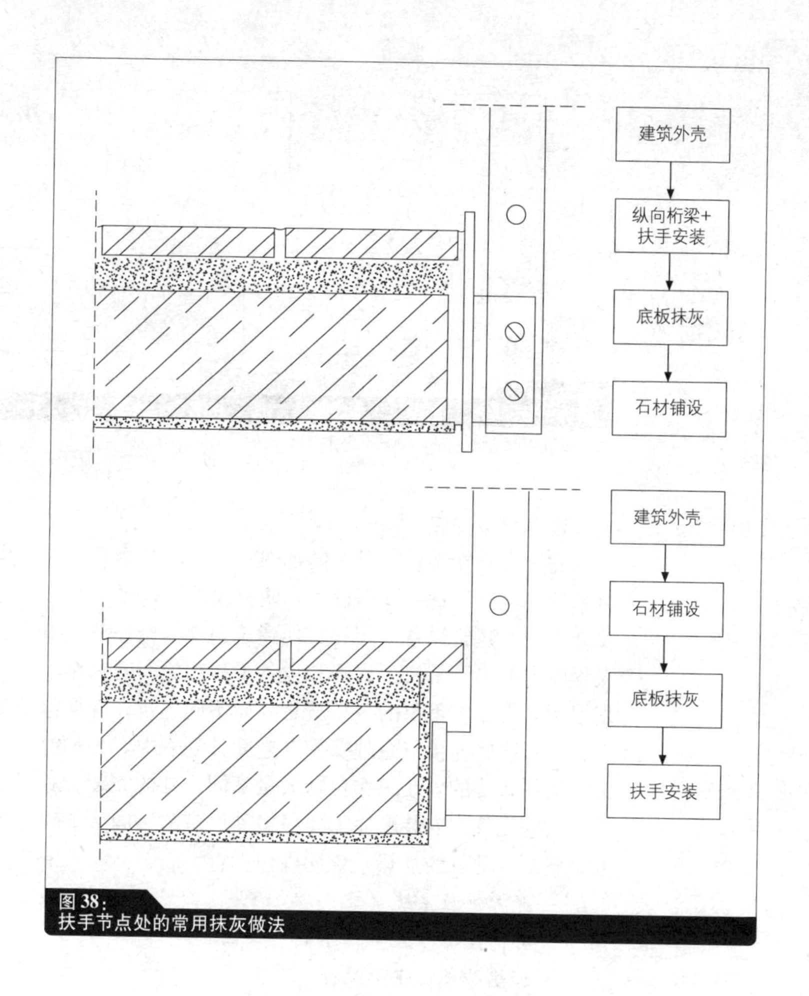

图 38：
扶手节点处的常用抹灰做法

墙面涂装与墙纸铺设

在进行墙面涂装或者墙纸铺设之前需要确保墙面处于干燥状态。因此，在进度计划编制时应该针对石膏等含有矿物质的表面材料留有足够的干燥时间。一般来说，墙面涂装或者墙纸铺设应该覆盖所有未铺设其他表面材料（比如墙面砖或者预制吊顶材料等）的区域。对于一些细小的工作，比如楼梯扶手、钢门框的油漆，电梯安装之前对于电梯井的防尘、抗油层的铺设以及钢结构防火、防腐涂料的涂装等，计划编制者也应该给予充分的考虑。在进度计划表中，应该对抹灰工作的后续工作加以相关说明。

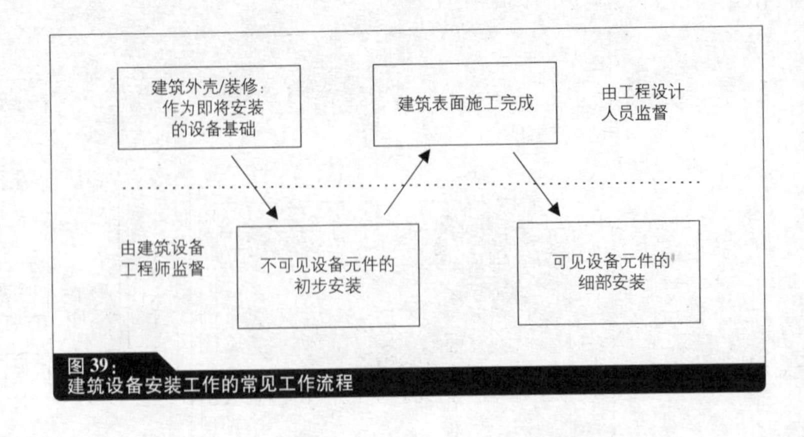

图 39：
建筑设备安装工作的常见工作流程

P61

建筑设备安装

建筑设备安装包括所有的加热系统、给排水系统、卫生设备、通风系统、电气设备、数据设备、防火设施、电梯以及其他特殊设施的安装工作。建筑设备安装和其他建筑内部装修任务之间的协调工作很大程度上取决于装修工程师和设备工程师之间的合作。因此，对于计划编制人员非常重要的一点就是弄懂不同建筑设备承包商之间的连接点，并将这些连接点合理安排在进度计划表中。〉参考图 39

加热系统安装

加热系统的安装工作包括大量不同元件的安装。对于某一特定工程，这些元件的安装顺序需要根据其设有的不同系统和分布管道的类型来确定。常见的加热系统元件包括：

— 能源供应系统（比如燃气管道、太阳能收集器以及水管等）；
— 存储设备（比如储水箱、储油箱等）；
— 加热站或者供热站；
— 建筑内部的主要管道和分支管道（管道安装）；

举例：

加热设备的安装是一个典型的交接点问题。在冬季施工时，为了确保涂装层和面层内部能够适当地干燥，需要安装加热设备。然而，在有些情况下为了涂装加热设备后面的墙体，需要再次移动加热设备的位置。

— 单个散热器的布置（散热器连接）；

— 散热器的安装。

进度计划编制人员需要对所有建筑内表面的加热系统元件的安装工作进行协调安排。对于冬季施工时用来加热施工现场的加热设备，可能需要提前安装部分加热设备，然后再进行临时移动以便于进行加热元件附近的面层施工（比如抹灰和刷漆等）。

卫生设备的安装

卫生设备与加热设备类似，都需要进行闭合（有时需要增压）回路和管网的安装。因此，卫生设备的安装流程与加热系统类似。在进行饮用水系统设计安装的时候，设计人员除了需要考虑建筑内部管道与本地供水管道相连之外，还需要清楚给水管、增压管在管道竖井或墙槽中的分布情况，以及与卫生间、厨房等房屋中出水口的连接情况。另外，还需要对下水管道进行相同的设计。

在初步安装之后进行的是细部安装（比如水槽、厕所、水龙头等）。这项工作一般安排在贴瓷砖和涂装工作完成之后，〉参考图 41 以避免可能的损坏或者失窃。对于浴缸、淋浴盆等外部需要贴瓷砖的设备，应该在贴砖工作开始之前进行安装。

常见不同工作之间需要协调的连接点如下：

— 底板内部的管道铺设（一般在建筑外壳施工阶段进行安装）；

— 墙体以及楼板开洞（在安装完成后，有些情况需要在建筑外壳施工阶段进行封堵）；

— 沿屋顶分布的通风管道（需要将下水管道与屋顶处的排风扇进行连接）；

— 上、下水管道与建筑的连接（需要与市政部门进行协调）；

— 回水水位以下区域水泵系统的安装；

— 饮用水加热设备（采用中央供水装置以及平行于饮用水管道的布置，或者在用水点进行分别安装）。

电气设备的安装

电气设备的安装同样可以分为初步安装和细部安装两个阶段。根据安装方法的不同，电缆线可以封闭在砂浆层内部或者直接暴露在室

提示：

有关饮用水和排水系统组件的信息可参考：本套丛书中的《水循环系统》（中国建筑工业出版社 2011 年 3 月出版，征订号：20133）

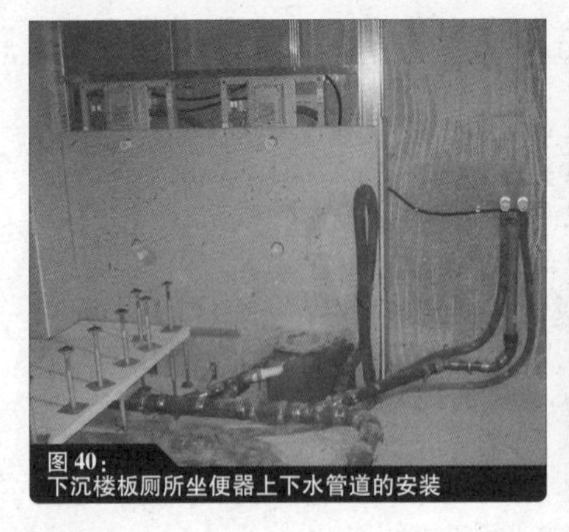

图 40：
下沉楼板厕所坐便器上下水管道的安装

图 41：
贴砖完成后可进行卫生设施的细部安装工作

内。对于住宅建筑，如果不想让电缆外露，则整个电气网路的施工必须在建筑外壳施工与抹灰工作之间完成。而对于工业建筑，一般采用外露安装的方式，其安装工作通常在所有墙面施工完成之后进行。但是，外露混凝土墙体是一个例外，电缆套管的安装应该在混凝土浇筑之前铺设完毕，以便于随后的电缆安装。〉参考图 42 电气系统安装的标准工作阶段如下：

— 建筑供电与主保险丝的安装（与电力部门进行协调）；
— 接地线（与建筑外壳施工工作进行协调）；
— 蓄电池与变压器安装（若有需要）；
— 室内主电缆以及各用电处分支电缆的安装；
— 电灯、插座、开关等的细部安装。

由于电气设备与结构构件的一体化程度越来越高，对于电气工作的进度计划安排也变得越来越复杂。为了避免开凿已装修完成的建筑表面进行附加安装，计划编制者应该对建筑构件的施工与电气设备安装之间的连接点进行系统考虑。常见的问题包括：

— 电炉、瞬时热水器、加热系统以及特殊结构构件的专用电路；
— 建筑外部照明系统；
— 应急照明系统；
— 火灾报警设备；
— 通风设备；
— 排烟排气设施（窗户、天窗、排烟口）；

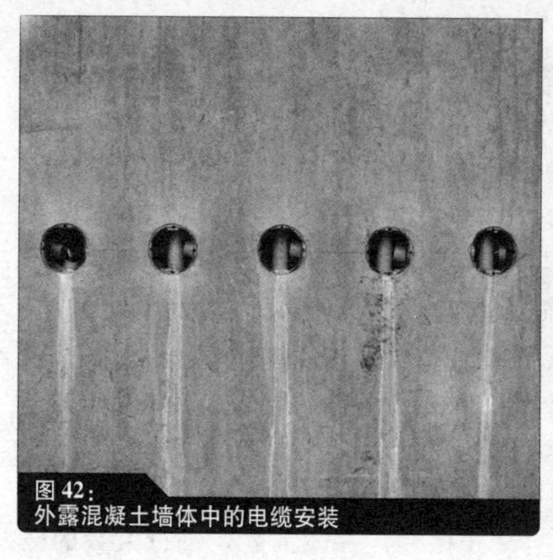
图 42：
外露混凝土墙体中的电缆安装

图 43：
楼板内部电缆分布示意

— 残疾人入口（开关、自动开关门系统）和入口监视器；
— 报警系统元件（外门、紧急出口、窗户防盗器以及监视摄像头等）；
— 立面控制（通风设备、太阳关以及眩光保护、大风及下雨报警器、顶灯等）。

数据系统

数据系统是电气设备安装工作中的一项特殊任务，同时也是特别复杂的一项任务，主要用于建筑的管理和通信。数据系统包括各种形式的电信和媒体技术，比如电话机、电视、电脑网络、机房等。

为了方便地进入需要管理建筑的工作区中，机房和布线室一般采用集中设置或者各层设置的形式。在编制进度计划时需要将楼板、墙体以及顶棚中的电缆线槽工作考虑在其中。

通风设备

在进行通风设备或者空调设备的安装工作安排时，还需要将进风管道和排风管道的布置考虑在其中。一般来说，不论通风管道采用了外露或者封闭形式，都会布置在管道竖井、楼板内部或者顶棚中。所以，计划编制者必须确保这些管道的安装工作能与相关的结构或者面层施工工作协调一致。在这方面，对于供给管道（包括冷却水管、电缆、进风管道）、建筑外表面防渗（由屋顶承包商或者立面装修承包商来进行开凿后的填补工作）以及防火分区（由建筑外壳施工承包商进行阻燃剂涂抹或者墙体承包商进行阻燃封闭）等工作的考虑非常关键。

在进度计划编制过程中，除了需要考虑通风设备与其他分布管道系统的连接点之外，还需要协调其他元件的细部安装工作，比如排风孔、格栅、挡风板以及面板等。细部安装工作一般在面层施工结束之后进行。

另外一个需要重点考虑的问题是，对于大型通风设备和空调设备安装，需要为设备的制造给定足够的准备时间。除了少数标准通风管道之外，大多数的管道、节点以及相关元件尺寸通常需要在建筑外壳施工过程中进行现场计算得到，而且必须采用专门的施工图来表达。只有在该图纸审查通过之后，才可进行管道构件的制作，通常制作周期需要若干星期。因此，通风设备的安装工作一般是由专业的公司进行承包，而且是在工程的较早阶段进行合同签订，以确保现场的安装工作能够按时完成。

运输设施

运输设施（比如电梯、自动扶梯等）一般都需要进行大量的电力安装。所以在进度计划编制时，需要考虑它们与楼板（电梯车厢内部地板，电梯门槛等）以及墙体（电梯内外门）覆层之间的连接点。

在大多数情况下，电梯的停靠点是根据安装在电梯井中的电梯轨道来确定的，而电梯轨道的安装则是在建筑外壳施工过程中进行的。所以，设计者需要尽早确定电梯生产厂家。当建筑外壳施工完成之后，将对电梯井尺寸进行精确测量，然后根据测量结果绘制电梯施工图。在预组装工作完成之后，具体的安装工作将分为若干步骤完成。首先，进行电梯承载系统的安装，然后是电梯车厢的安装，最后是电子控制系统的安装以及与电气设备的连接。

在进度计划安排时，还需要考虑安装的电梯是否会在随后的施工过程中用来运输建筑材料。不过，通常情况下不建议将电梯用来运输建筑材料，以避免对电梯车厢造成不可预见的破坏。因此，进度计划编制者经常有意在施工工作的较晚阶段来完成电梯的安装工作。

P65

收尾工作

竣工前的收尾工作

除了各个工程参与方已经指定的后续工作之外，在施工过程的最后阶段还有一系列的收尾工作需要完成。主要包括：

— 清扫工作，在所有工作完成之后，交付使用之前；

— 闭锁系统的安装和移交（最终闭锁系统的安装并交付给用户）；

— 完成室外空间的施工（通道、花园及草坪设计，停车场，标志，室外照明系统及其他设施的安装）。

法律要求的验收

一般来说，比较有效的做法是在工程结束之前留有一段时间，用来纠正施工中的错误和迎接竣工验收。这些工作将需消耗一段时间，而且需要在建筑交付使用之前完成。

竣工验收内容包括合同中所规定的内容和法定的相关内容。对于后者，建筑管理委员会将对建成的房屋是否满足建造规定，以及是否满足使用要求进行验收。这些验收内容还包括设备的安装，比如防火设施、加热系统以及空调系统等，而且这些验收工作有时还需要由外单位专家来完成。

P67

施工进度计划表的使用

即使施工进度计划表能够清楚地表示出一个工程项目的所有细节安排，但进度计划表并不是一个静态的表格，不会从制订完成直到施工完成的期间保持一成不变。在施工过程中，随着现场施工条件的变化，需要针对给定工期的施工项目进行必要的计划调整。所以说，施工进度计划表应该作为伴随整个施工过程的工具来看待。

P67

施工进度计划表的更新与调整

计划与现实

现场的施工条件一般会与编制进度计划表时所设想的条件存在差异，而且还会因为各种原因造成施工的中断或者结构的变更。〉参考"施工进度计划表的使用"一章中"施工过程的中断"部分内容而进度计划表通常是打印出来张贴在施工现场，所以很快就会过时，不再对施工任务有约束作用。因此，必须对进度计划表进行更新。施工进度计划表并不需要作为一系列强制任务而根据实际施工情况来不断更新，而是应该作为施工人员管理、组织或者必要情况下调整实际施工过程的一个日常工具来使用。

施工进度计划表的框架层次

在制定施工进度计划表的时候，编制人员应该确保进度计划表在施工过程中的有效性，同时便于更新和添加。对于大型项目，由于施工任务的数量和复杂性，施工进度计划表通常非常复杂。在这种情况下，单独的施工任务应该在一个清晰的汇总任务框架下分层次进行计划安排。〉参考图 44 这种方法不仅能够清晰地表达施工阶段和施工流程的细节，同时还能够表达出整个工程项目的总体截止期限。

面向用户的进度计划表

面向用户的进度计划表能够帮助施工人员从整体上更加容易地理解整个进度计划安排。根据具体用户的不同，进度计划表可以采用不

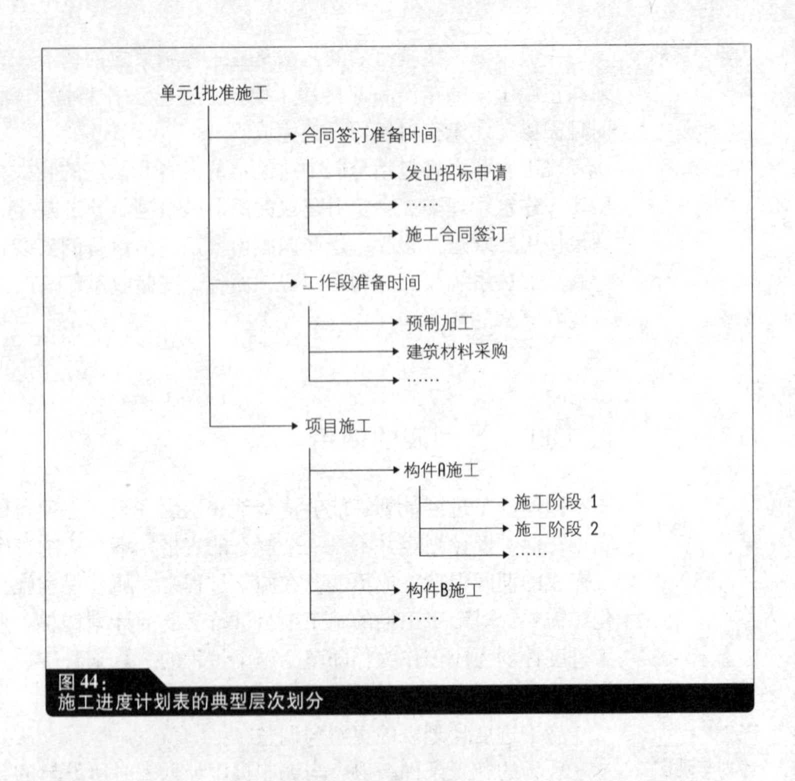

图 44：
施工进度计划表的典型层次划分

同的形式。进度计划表的常用形式包括：

— 概述——针对项目经理和业主：包括主要汇总工作，不包括每个单独施工任务；
— 设计和合同签订截止期——针对设计单位：包括所有的单独设计任务和设计合同签订准备时间，不包括施工任务；
— 施工进度计划表——针对现场经理：不包括设计任务和设计合同签订准备时间，包括所有的施工合同签订准备时间和所有的施工任务；
— 施工进度计划指南——针对每个相关的工程参与人员：仅仅显示每个工程参与人员各自相关的工作。

对于每个单独的施工小组，相应的进度计划表中仅表示出与之相关的任务内容，能够使小组成员更加明确各自的任务计划和实施过程。

不同的进度计划表表示形式的一个重要基本原则是它们根据同一个整体进度计划得来。假如不同的进度计划表按照平行的方式进行编

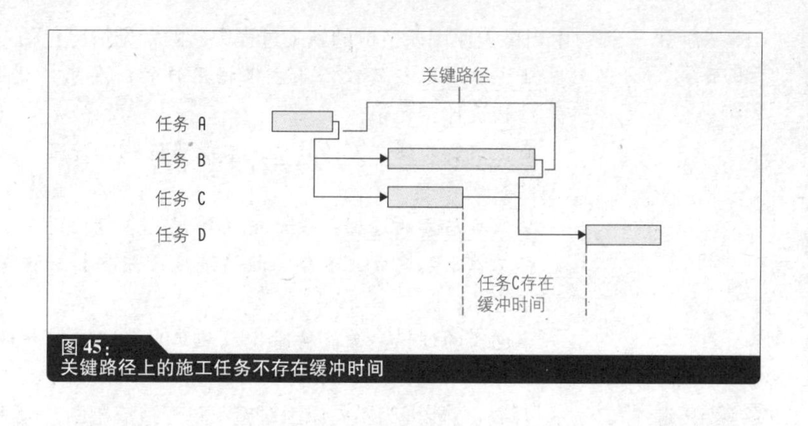

图 45：
关键路径上的施工任务不存在缓冲时间

制，那么不同的使用者以及他们之间的相互影响会导致实际施工工作的一致性难以得到保证。而将所有的修改情况整合到同一处，然后按照上文所介绍的方法分层次传递到相关的工程参与人员手中，则可以使施工进度计划表在整个施工过程中的作用更加显著。

修改变更的整合

除了分层次对进度计划表进行编制之外，通过建立不同的任务组之间清晰的相互联系关系也可以提高进度计划表的实用性。为了保证修改变更在不同进度计划表中的一致性，需要将所有的任务整合到同一个地方，然后确保整表能够自动更新。同时，需要注意的一点是保证所有的推迟和修改内容不会影响到工程项目的竣工时间。

关键路径

有些情况下，仅存在一条路径贯穿一个工程的始终。当这条路径受到影响推迟之后，整个工程的工期将受到直接的影响，这条路径就是所谓的关键路径。此时对于关键路径之外的其他施工任务，则存在一段缓冲时间来避免对关键路径造成影响。〉参考图 45

缓冲时间

对于不在关键路径上的施工任务均存在缓冲时间，其长度可以采用现代编制软件来计算。编制人员在面对可能的工期延迟时，可以利用任务的缓冲时间来对相关任务进行灵活调整，同时还可以通过缓冲时间的计算确定允许施工单位的额外工期的长度。

P69

施工过程的中断

编制人员针对施工进度计划表所进行的必要的修改大多数情况下是由施工过程的中断所导致的。导致工程中断的原因可能是当事人（包括业主、设计方以及签订合同的施工单位）、承包商或者第三方。

当事人所引起的施工中断

由当事人原因所造成的施工过程中断的常见情况包括：

— 业主要求进行修改：业主根据用户的新要求所进行的修改，结构设计变更等；

— 业主未能起到应有的作用：未能获得施工许可，未支付施工款等；

— 工程参与方的过错：设计中的错误、未能按时完成设计、招标工作未能完成、不合实际的进度计划安排、施工监督过程不力等；

— 承包商的过错：未能按时完成施工准备工作，导致不能按时为施工单位提供施工场地。

承包商所造成的施工中断

承包商也可能因为不同原因对施工过程造成中断。其中最糟糕的情况是，承包商因为无力偿还债务而宣告破产。此时，业主不得不重新寻找其他承包商并重签合同以完成剩下的施工工作，但工期将遭到很大程度的推迟。另外，如果施工单位签订的施工任务太多，那么很可能在人员安排方面无法满足施工合同中的要求，这样也会造成施工工作的推迟。另外，比如罢工、流感爆发等原因，也会大大减少施工单位在施工工地的人员数量安排。

人员数量规划

在进行人员数量安排时也会经常遇到不同的问题。施工单位在某个施工场地的人员安排一般是按照固定的周期（比如，以星期为单位）进行改变的，同时不同的施工人员可以在不同的工地之间交叉使用。施工单位一般不会针对某个施工工地按天改变施工人员的数量，如果在施工进度计划中每天对于施工人员的需求都不相同，那么将很有可能会造成施工过程的中断。〉参考图 46

注释：

通过不同颜色标注、柱状图以及不同的填充方式可以对不同的施工任务、节点、汇总任务以及整个施工场地进行区分，从而增加进度计划表的可读性。通过这种方式，能够非常容易地对不同的工作段以及施工过程进行区分。任务的自动标注功能也是一种比较有效的方法。

注释：

对缓冲时间的计算还具有另外一个好处。如果施工过程中依次进行的两个施工任务不存在相互联系，那么计算所得的缓冲时间能够相应地延长，直到整个工程的截止期。进度计划编制人员可以非常容易地通过检查对较长的缓冲时间的检查，来核实是否忽略了任务之间可能存在的依赖关系。

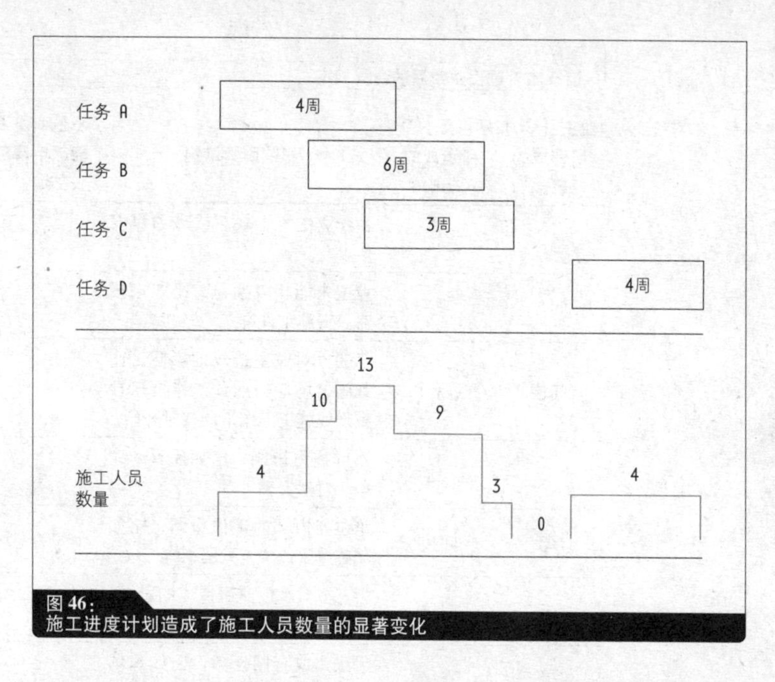

图 46：
施工进度计划造成了施工人员数量的显著变化

因此，为了避免随后可能带来的影响。在编制进度计划表时应该尽可能确保不同工作段之间的施工人员能够保持一个相对不变的状态。

〉✎

第三方所造成的施工中断

除业主和承包商之外，第三方也可能介入施工过程中并造成施工过程的中断。造成施工中断的原因可能来自政府部门、罢工、业主或者承包商的失窃以及不可抗力等。如果由于第三方的原因造成了业主所提供施工条件的改变，那么承包商有权要求在原有施工工期的基础上进行适当的增加。如果由于第三方的原因增加了承包商的风险，此时承包商仍然需要按照合同规定按时履行自己的义务。如果遭受了不可抗力的影响，比如风暴、洪水等，施工单位一般可以被批准将工期进行相应延长。〉参考表 2

注释：

为了确保施工过程中的施工人员保持一个相对平稳的状态，在编制施工进度计划时不仅需要考虑不同工段中的同一工作任务之间的关系，还需要考虑同一工段中不同工作任务之间的相互关系。所以，进度计划编制者应该预先将同一工作段的施工人员划分为若干可以连续工作的施工小组。

表 2：
造成施工过程中断后的合同规定

业主（CL）所造成的影响	承包商（CN）所造成的影响	影响因素示例	承包商是否能够申请延长工期？	是否能够申请向业主进行索赔？
直接	无	由于业主资金缺乏或者项目发生变化	是	是
间接	无	业主未能积极参与，比如未按时获得施工许可	是	是
间接	无	第三方对业主造成的影响，比如施工场地的准备工作推迟导致现场施工工作不能按时开始	是	是
无	无	不可抗力影响，比如风暴、战争、自然灾害等	是	否
无	间接	承包商内部的工作中断，比如流感爆发或者工人罢工等	否	否
无	间接	第三方对承包商所造成的影响，比如设备失窃等	否	否
无	直接	无法完成合同任务，比如现场施工人员数量太少	否	否

天气的影响

即使是大型工程项目，冬季以及一年内其他时段里不利的天气条件也可能会造成施工过程的延迟。对于在冬季条件下以及节假日所造成的工作效率的降低，虽然可以在较长的任务缓冲时间和持续时间内进行补偿，但也不可能对冬季的天气进行准确预测。对于某些地区，由于霜冻等天气条件的影响，可能会导致施工工地处于长时间的停工状态，而通过尽早进行采暖系统或者现场加热系统的安装，则可以避免这种情况的发生。然而对于一些建筑材料，比如预拌混凝土、浇筑式沥青以及预拌砂浆等，即使建筑内部的温度足够高，但如果室外温度太低也无法进行施工工作，在进行施工进度计划安排时必须加以注意。

施工中断的分类

施工进度计划的中断可以大致分为以下三种类型：〉参考图 47

— 推迟完成：任务开始时间推迟，但在预定的任务持续时间内完成；
— 施工时间延长：任务所需的施工时间比计划时间长；
— 施工顺序改变：打破施工过程或者相互联系的施工任务的原有计划安排，进行顺序上的重新调整。

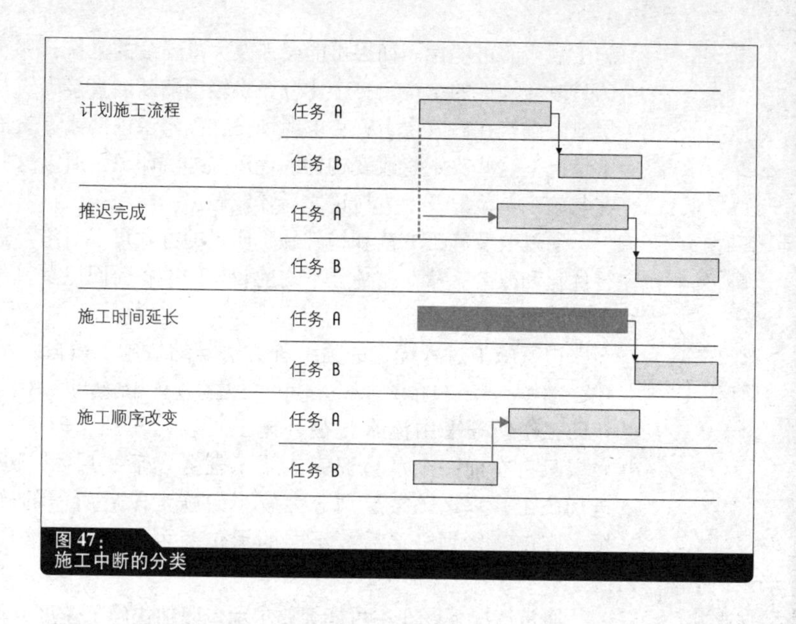

图 47：
施工中断的分类

P73

施工中断的应对

万一由于施工顺序的改变对施工关键路径造成了影响，或者可能导致整个工程无法在规定的截止期内完成，进度计划编制人员应该尽可能在预定的施工工期内解决这些问题。可能的影响因素包括：

— 必要任务依赖关系的检查；

— 施工过程的顺序改变；

— 施工方法与施工质量的改变；

— 缩短施工阶段；

— 加快施工速度。

必要任务依赖关系

并不是所有的任务依赖关系都是必要的，即使是一些看起来比较合理的依赖关系。若对工作任务进行充分的考虑，而且其与其他构件的连接可以在施工后再进行施工，那么有些任务有时可以采用逆序的方式进行施工。

进度计划编制人员手续应该采用从前往后的方式来（比如砂浆铺设→涂装）确认任务之间的相互关系是不是必需的，或者考虑不同的任务是否可以采用其他的顺序进行组织。如果遇到问题，最好与相应的施工单位进行讨论。

施工顺序的改变

如果对于施工过程的影响较大，造成了对必需的任务依赖关系的影响。此时，为了满足工期的要求，进度计划编制人员需要考虑改变施工顺序的可能性。如果可能会涉及大量连续的工作，那么比较可行的是改变构件的结构。打个比方，将轻型隔断墙直接砌筑在楼板结构层或者砌筑在砂浆层上；将电缆铺设在砂浆层内部或者表面。但在通常情况下这些决定将改变构件的性质和外观印象，所以这些决定必须在与施工单位以及甲方讨论之后才能做出。

施工方法与施工质量

通过改变施工方法和构件施工质量的方式可以对施工进度计划进行优化和改善，从而避免任务之前较长的准备时间以及较长的养护和干燥时间。

缩短施工阶段

进行施工过程优化的另一种方法是缩短施工阶段。根据前文所述，〉参考“进度计划的创建”一章中“工程进度计划的结构”部分内容对背靠背的工作任务采用流水化循环施工可以缩短施工时间。如果某一楼层只有在前一个承包商完成施工任务之后，后一个承包商才能进场施工，那么该楼层划分为较小的施工段后所需的施工时间更短。在施工段划分之后，不同的承包商可以在同一楼层同时进行施工。

加速施工

通常情况下，业主可能会要求承包商加快施工速度，但这种情况与施工延迟所需的加快施工情况有所不同。如果是因为承包商的责任需要加快施工速度，那么承包商必须采用所有可能的措施——包括要求工人加班工作以及雇佣更多的工人等——来满足合同中的截止期，而且必须在不考虑施工成本的前提下来进行。但是，如果是业主或者项目经理要求施工延迟的无责任施工单位加快施工速度，那么业主必须承担额外的施工成本。

避免额外成本的发生

由于上文所述的一些加快施工速度的措施会产生额外的施工成本，所以业主必须参与做决策的过程之中。首先，业主必须确定其能够为确保建筑按期完工所能够动用的资金数额。

举例：

对于一些会产生大量灰尘的工序，比如抹灰、找平以及水磨石施工等，应该安排在地毯铺设、墙面涂装等容易被弄脏的工序之前。但对于楼梯而言，其踏步地毯的铺设只要在踏步完成并打扫干净之后即可进行，并不是必须在完成所有石材踏步的施工之后进行。

举例：

在工期比较紧张的时候，如果计划编制人员想要避免水泥砂浆找平层的养护时间和干燥时间，那么可以采用铺设沥青砂浆或者其他干式找平层等替代的方法，这样在第二天就可以上人工作。然而，此时需要注意的一点是这些替代方法比水泥砂浆的成本高出许多。

即使是在编制进度计划的最初阶段，也建议编制人员将工程施工的延迟时间考虑在整个进度计划之中。在施工过程中几乎总会出现各种问题——准备工作不够充分、材料采购滞后、失窃等——而且这些问题必须得以解决。在进度计划中设置适当的缓冲时间是能够有效确保工程按期完成的一个重要方法。如果在进度计划表中没有安排缓冲时间，那么意味着该工程几乎无法按期完工。

考虑替代方法

此外，在工程的早期阶段，进度计划编制人员应该明白在不违反合同和增加额外成本的前提下允许采用替代施工方法（比如，现场浇筑混凝土还是采用预制方式，石膏还是石膏板，水泥砂浆找平层还是干式找平层，选用封闭门框还是采用灰浆抹平的方式）的最晚时间点。相应的工期应该及时地进行分析，并在与业主共同讨论后再做出相应的决策。

P75

作为过程文件的施工进度计划表

施工进度计划表不仅是进行施工过程协调的一种方法，同时还是一项工程的重要证明文件。由于施工进度计划表涵盖了一项工程的设计和施工阶段，所以它可以作为证明或者反驳施工中断的证据。当业主和施工单位之间存在不确定的索赔（比如损坏赔偿金等）时，进度计划表则显得非常重要。另外，已经竣工的工程的施工进度计划表则是制定未来工程进度计划表的重要参考，同时也是进度计划编制人员从施工过程中所获得的重要经验。

实际截止期与工程中断的记录

从此方面出发，进度计划编制的主要任务是记录任务的实际持续时间，并将其与编制时的估算时间进行对比。施工过程的中断以及造成中断的原因应该加以记录。一种记录的方法是将施工现场所发生的事情通过手写的方式记录在当时的进度计划表中。将现场施工进度计划表定期复印入档，则为进度计划表的更新提供基础。理想情况下，项目经理可以将截止期直接填入进度计划编制程序中来完成自动实时更新。但是，在每次更新之后，需要及时将先前的进度计划版本进行存档。

P77

结语

复杂项目的工程施工需要进行大量的组织和协调工作。如果没有有效的进度计划，将无法对大型施工项目进行有效的时间管理。对于负责协调工作的建筑师以及现场项目经理，非常重要的一点是需要提前对所有的设计和施工过程进行组织协调，以便有效地控制整个局面。如果负责建筑师和项目经理只是在情况发生之后做出被动的响应，而无法对施工过程进行主动控制，仅由相关参与人员进行自发的尝试，那么必将导致施工中断、无法协调、相互影响和施工延迟等情况发生。

尽管如此，对工程设计和施工过程的管理并不是为工程参与人员提供一项必须严格遵守的具有固定截止期的进度计划表。进度计划编制专家应该考虑到所有工程参与人员所关心的问题，并将它们整合到管理过程中，以找到所有参与人员能够贯彻实施的解决方法。

进度计划表不仅是建筑业主和建筑师之间所签订的合同服务内容，同时也是一种对设计和施工过程进行日常管理的有效工具。制定一项有效并且实施性强的进度计划表需要付出大量的努力。如果编制人员能够提前将整个设计和施工过程进行全盘考虑，那么随后的协调工作和冲突的解决工作将会更加容易处理。建筑师在提前进行施工过程顺序安排时投入越多，项目经理现场的工作将越容易。

附录

参考文献

Bert Bielefeld, Thomas Feuerabend: *Baukosten- und Terminplanung*, Birkhäuser Verlag, Basel 2007

Tim Brandt, Sebastian Th. Franssen: *Basics Tendering*, Birkhäuser Verlag, Basel 2007

Chartered Institute of Building (ed.): *Planning and Programming in Construction*, Chartered Institute of Building, London 1991

Wilfried Helbig, Ullrich Bauch: *Baustellenorganisation*, Rudolf-Müller-Verlag, Cologne 2004

Hartmut Klein: *Basics Project Planning*, Birkhäuser Verlag, Basel 2008

Werner Langen, Karl-Heinz Schiffers: *Bauplanung und Bauausführung*, Werner Verlag, Neuwied 2005

Richard H. Neale, David E. Neale: *Construction Planning*, Telford, London 1989

Jay S. Newitt: *Construction Scheduling. Principles and Practices*, Pearson Prentice Hall, Upper Saddle River, NJ, 2009

Lars-Phillip Rusch: *Basics Site Management*, Birkhäuser Verlag, Basel 2008

Sandra Christensen Weber: *Scheduling Construction Projects. Principles and Practices*, Pearson Prentice Hall, Upper Saddle River, NJ, 2005

Falk Würfele, Bert Bielefeld, Mike Gralla: *Bauobjektüberwachung*, Vieweg Verlag, Wiesbaden 2007

P79

进度计划编制过程中所需的信息

表 3：
工程最初阶段进度计划编制所需要的标准信息

信息提供方	信息需求方	信息内容
结构工程师	建筑师	— 结构体系和结构材料 — 结构构件的所有尺寸信息
建筑设备工程师	建筑师	— 为满足使用功能所需要的标准设备 — 杂物间和配线室的位置 — 主电缆的线路，和主要分支电缆的线路 — 设备和线路的初设尺寸
建筑师	结构工程师	— 场地平面图，建筑形式，楼层高度 — 柱子的最大尺寸和最常用尺寸
建筑师	建筑设备工程师	— 场地平面图，建筑形式和尺寸 — 建筑用途和用户数量（比如，若用做办公室则需确定雇员的数量） — 建筑设备安装需求 — 建筑平面图

表 4：
工程设计阶段主要设计人员所需要的标准信息

信息提供方	信息需求方	信息内容
结构工程师	建筑师	— 结构承载构件的主要轴线和二级轴线 — 结构构件的初设尺寸
建筑设备工程师	建筑师	— 设备和管线的初设尺寸 — 建筑设备的开洞要求 — 成本估算
建筑师	结构工程师	— 尺寸设计深化图（平面图和剖面图），用来申请建筑工程规划许可证
建筑设备工程师	结构工程师	— 主要管线位置，设备安装位置以及设备荷载
建筑师	建筑设备工程师	— 最终深化设计图（平面图、剖面图以及效果图）
结构工程师	建筑设备工程师	— 承载结构设计（梁、柱、承载墙等） — 承载构件的开洞和开槽情况

表 5：
施工准备阶段主要设计人员所需要的标准信息

信息提供方	信息需求方	信息内容
结构工程师	建筑师	— 模板图 — 配筋图 — 节点详图 — 材料表
建筑设备工程师	建筑师	— 电气设计图、通风设计图、暖通设计图、卫生设备设计图 — 建筑设备所需的墙面开洞、开槽设计图 — 招标文件，比如招标邀请中主要大纲 — 其他专业设计专家所要求的条件
建筑师	结构工程师	— 更新尺寸之后的平面图和剖面图 — 设计施工图，设计详图和设计说明 — 建筑工程规划许可证以及前期项目说明中的相关规定
建筑设备工程师	结构工程师	— 主要管线位置，设备安装位置以及设备荷载 — 开洞和开槽情况
建筑师	建筑设备工程师	— 政府部门所批准的设计方案或者相关说明 — 设计说明 — 设计施工图
结构工程师	建筑设备工程师	— 模板图、钢结构设计图以及木结构设计图 — 墙体开洞处的钢筋位置

时间定额

表 6：
任务持续时间估算所采用的时间定额示例

任务内容	时间定额	定额单位
施工场地准备		
起重机架起	10 ~ 50	h/unit
钢结构围墙	0. 2 ~ 0. 4	h/m
管线连接（电气管道，水管）	0. 2 ~ 0. 5	h/m
土方开挖		
建筑基坑开挖	0. 01 ~ 0. 05	h/m^3
采用挖掘机进行独立基础开挖（包括土方运输）	0. 05 ~ 0. 3	h/m^3
独立基础人工开挖	1. 0 ~ 2. 0	h/m^3
混凝土工程		
建筑外壳施工估算（毛体积 700 ~ 1000m^3，工人 3 ~ 5 个）	0. 8 ~ 1. 2	h/m^3 GV
粘结层，无配筋，厚度 5cm	0. 2	h/m^2
底板，现浇混凝土，厚度 20cm	2. 0	h/m^2
楼板，现浇混凝土，厚度 20cm	1. 6	h/m^2
预制或者部分预制混凝土楼板	0. 4 ~ 0. 9	h/m^2
整体预制结构	0. 3 ~ 0. 7	h/t
构件混凝土浇筑（不包括模板工程和钢筋绑扎）	0. 4 ~ 0. 5	h/m^3
墙体混凝土浇筑（不包括模板工程和钢筋绑扎）	1. 0 ~ 1. 5	h/m^3
柱子混凝土浇筑（不包括模板工程和钢筋绑扎）	1. 5 ~ 2. 0	h/m^3
楼梯现场混凝土浇筑（不包括模板工程和钢筋绑扎）	3. 0	h/unit
大面积模板工程	0. 6 ~ 1. 0	h/m^2
单构件模板工程	1. 0 ~ 2. 0	h/m^2
钢筋绑扎	12 ~ 24	h/t
不同形式的密封工作	0. 25 ~ 0. 40	h/m^2
脚手架（安装与拆除）	0. 1 ~ 0. 3	h/m^2
砌体工程		
承重砌体墙	1. 2 ~ 1. 6	h/m^3
非承重内墙	0. 8 ~ 1. 2	h/m^3

续表

任务内容	时间定额	定额单位
木结构工程		
屋椽，包括连接和安装（从屋顶区域）	0.5～0.7	h/m^2
屋面工程		
平屋顶（砂砾层），包括非隔热屋顶的全安装工作	0.5～0.7	h/m^2
铺瓦坡屋顶	1.0～1.2	h/m^2
金属屋面	1.3～1.5	h/m^2
外墙覆盖		
金属立面覆盖	1.0～1.3	h/m^2
护面砖	1.1～1.5	h/m^3
复合隔热系统	0.6～0.8	h/m^2
预制混凝土立面材料安装	0.5～0.7	h/m^2
外墙天然石材、岩石覆盖	0.5～0.8	h/m^2
窗户结构		
独立窗户安装	1.5～2.5	h/unit
住宅卷闸门安装	0.6～1.5	h/unit
屋面窗户	2.5～3.5	h/unit
室内窗台	0.3～0.5	h/m
抹灰		
外部抹灰	0.5～0.7	h/m^2
机器内部抹灰	0.2～0.4	h/m^2
人工内部抹灰	0.3～0.6	h/m^2
顶棚抹灰	0.3～0.4	h/m^2
找平层		
水泥砂浆和脱水材料找平层（无面层、隔声层等）	0.1～0.3	h/m^2
石油沥青砂胶找平层	0.3～0.5	h/m^2
浮隔地板找平层（含隔声层）	0.6～1.0	h/m^2
水磨石找平层，抛光	2.0～2.5	h/m^2
干式结构		
石膏板干式墙	0.2～0.5	h/m^2

续表

任务内容	时间定额	定额单位
预制墙（面），单面墙体，含墙体基础	0.7~0.8	h/m^2
含斜撑顶棚铺设	0.3~0.5	h/m^2
悬挂顶棚结构	0.6~1.1	h/m^2
石膏龙骨墙体，单面	0.4~0.8	h/m^2
石膏龙骨墙体，双面	0.6~1.5	h/m^2
门		
钢结构门框+门扇安装	1.9~2.5	h/m^2
木门安装	1.0~1.5	h/m^3
外门	2.5~4.5	h/m^2
瓷砖，铺路石与琢石		
地板瓷砖	0.5~1.8	h/m^2
墙面砖	1.3~2.5	h/m^2
天然以及混凝土铺路石	0.8~1.2	h/m
瓷砖和天然石踢脚板	0.3~0.4	h/m
地面		
找平层，注入孔	0.05~0.2	h/m^2
聚氯乙烯，油地毡和地面卷材	0.3~0.6	h/m^2
找平层上铺设针毡或者地毯	0.1~0.4	h/m^2
踢脚板	0.1~0.2	h/m^2
拼花地板，含表面处理	1.2~1.8	h/m^2
打磨拼花地板，含表面处理	0.2~0.3	h/m^2
天然石地板	0.9~1.2	h/m^2
楼梯踏步铺设	0.5~0.7	h/m^2
墙面涂装和墙纸		
腻子	0.1~0.2	h/m^2
标准墙纸（薄墙纸，浮凸墙纸等）	0.1~0.4	h/m^2
特殊墙纸（天鹅绒，纺织品，墙体画像等）	0.3~0.8	h/m^2
内墙涂装，单层	0.05~0.2	h/m^2
内墙涂装，三层	0.2~0.5	h/m^2

续表

任务内容	时间定额	定额单位
外墙抹灰，涂装	0.2~0.8	h/m^2
窗户涂装，每层	0.2~0.6	h/m^2
金属表面涂装，所需所有涂层（门，金属墙体等）	0.3~0.6	h/m^2
金属构件涂装，所需所有涂层（框架，金属覆面等）	0.6~1.0	h/m^2
金属扶手涂装	0.1~0.3	h/m
电气工程		
所有电器工程安装估算（毛体积700~1000m^3，工人2~3个）	0.2~0.4	h/m^3 GV
线盘及电缆线安装	0.3~0.5	h/m
灯具安装	0.3~0.8	h/unit
次级配电盘安装	0.5~1.0	h/unit
细部安装，开关，插座等	0.02~0.05	h/unit
加热系统，水泵和卫生设备安装		
加热系统安装估算（毛体积700~1000m^3，工人2~3个）	0.1~0.3	h/m^3 GV
燃气系统、给排水系统安装估算（毛体积700~1000m^3，工人2~3个）	0.15~0.4	h/m^3 GV
管道线路安装估算	0.4~0.8	h/m
雨水、废水管道安装	0.10~0.50	h/m
卫生设施的细部安装	0.3~1.0	h/unit

P85

作者简介

伯特·比勒费尔德，工学博士，德国多特蒙德自由执业建筑师，Aedis Paomanage 工程管理公司工程管理主管，在德国锡根大学（University of Siegen）教授工程经济学和工程管理，在多个建筑理事会和协会中进行讲授。